D1759120

for Thelma McCorquodale

botanic gardens

A Living History

black dog
publishing

Contents

Foreword

This book charts the history of botanic gardens from their beginnings in the sixteenth century at the Universities of Padua and Pisa to the most innovative gardens of today. With text from the curators and directors of some of the world's foremost botanic gardens interspersed with over 80 captivating profiles of some of the world's most important institutions, *Botanic Gardens: A Living History* is the first comprehensive look at these increasingly valuable organisations.

Four rigorous essays investigate the history and future of botanic gardens as institutions of conservation and recreation. A wide range of expert voices chart the development of the botanic garden, including Rosie Atkins (founder of *Gardens Illustrated* magazine), Mike Maunder (Director of the Fairchild Tropical Botanic Garden), John Parker (Director of Cambridge University Botanic Garden), Holly Shimizu (Director of the United States Botanic Garden), Gregory Long (President of the New York Botanical Garden) and Margaret Stevens (Founder and President of the Society of Botanical Artists).

Interleaved between these pioneering works there are profiles of world-renowned botanic gardens organised alphabetically based on the countries in which they are found. While it is impossible to include all of the world's botanic gardens in one volume, the gardens profiled are among the most noteworthy, and were chosen for their special contribution to the past and present of these most important institutions.

Dioscorides Ibn al-Baytar Materia Medica ...ltimate in Materia Medica ...dicaments and Nutritiona... Items Galen Luisa Tongiorgi Tomas... Wunderkammer Leonardo da Vinc... Albrecht Düre... Jacopo Ligozz... Luca Ghi... University of Pis... Carolus Clusiu... Hortus Botanicus Leide...

The History of the Botanic Garden

The History of the Botanic Garden

Elizabeth Barlow Rogers

The Botanic Garden in the Sixteenth and Seventeenth Centuries

The botanic garden is generally considered a Renaissance institution because of the establishment in 1534 of gardens in Pisa and Padua specifically dedicated to the study of plants. However, these and other sixteenth and early seventeenth century botanic gardens, including those at Breslau, Heidelberg, Kassel, Leiden, Leipzig, and Montpellier, did not spring into existence purely as a result of the great intellectual ferment of the times. Rather, they have their roots in the herbal manuscripts of antiquity, most notably the *De Materia Medico* of the first century CE Roman physician Dioscorides. This long-consulted reference work, which combines a brief discussion of a plant's physical characteristics with remarks about its remedial properties for specific diseases or injuries, was copied many times and remained the authoritative text on most known plant species well into the seventeenth century.

Discorides, first century CE Roman physician who wrote *De Materia Medica*, a botanical text still referred to today.

Many Dioscorides-based herbals perpetuated the knowledge and application of 'simples', as herbs were then called, and the collection and study of pharmacological herbs remained the focus of apothecaries and physicians both in Europe and the Islamic world. In the thirteenth century Ibn al-Baytar, 1179–1248, the most famous Arabic physician and botanist in Andalusian Spain, wrote two important books, *The Ultimate in Materia Medico* and *Simple Medicaments and Nutritional Items,* both of which were based on his own personal observations of some 1,400 plants, as well as knowledge derived from Dioscorides and the Greek physician Galen. As early as the tenth century, exotic collections were planted in Andalusian experimental gardens. Furthermore, Christian medieval art depicted images of gardens containing ornamental flowers that had symbolic value, the highest ranked being the rose and the lily, both emblematic of the Virgin Mary. And monasteries also contained gardens with collections of medicinal herbs. However, until Renaissance humanism revived a comprehensive and categorical Aristotelian approach to natural history there was little impetus to create botanic gardens as ordered collections of plants.

The earliest illustrations of pomology—the cultivation of fruit—drew on Greek mythology and depicted fantastically-sized fruit floating over garden or village scenes.

Factors besides pedagogy influenced the design of the earliest botanical gardens, and their layouts sometimes incorporated astrological, cosmological, and religious notions. As the botanical scholar Luisa Tongiorgi Tomasi points out, the geometric arrangement of Renaissance botanic gardens according to astrologically resonant forms—circles, squares, triangles—was intended to channel the positive energy radiating from the planets and stars into objects on earth, thereby increasing the healing power of the gardens' simples. The cardinal directions, cosmologically significant in all cultures, also influenced pre-Enlightenment botanic garden plans. In addition, early botanic garden designs embodied the biblical concept of paradise as an enclosed, geometrically ordered quadripartite space with four dividing paths symbolising the description in Genesis 2:10: "A river went out of Eden to water the garden; and from thence it was parted and became into four heads." Seen in this light, the arrangement of plants collected from the four corners of the earth in botanic gardens was intended to indicate a re-gathering of the paradisiacal bounty of Eden that was scattered at the time of the Fall. These early botanic gardens also should be understood as outgrowths of the gardens of princes and other wealthy individuals whose collections of rare plants were outdoor extensions of their *wunderkammer* cabinets containing all manner of exotica, both natural and man-made. The concept of the botanic garden as a kind of ethnographic and natural history museum can be traced back to this period.

Libraries were, as they remain today, essential elements of botanic gardens. The invention of the printing press in the fifteenth century greatly increased the opportunity for libraries to expand their collections, thereby extending the distribution of plant knowledge in general. However, the wood-block prints in Renaissance herbals retained the diagrammatic character of illustrations found in earlier illuminated manuscripts, and the close observation of actual plant forms, such as one sees in the drawings of Leonardo da Vinci, 1452–1519, and in the engravings of Albrecht Dürer, 1471–1528, did not become the norm in botanic illustrations until the second half of the sixteenth century when the second Grand Duke of Tuscany, Francesco I, 1541–1587, commissioned Jacopo Ligozzi, 1547–1626, to make watercolour drawings of the pineapple, fig, iris, and other exotic plant specimens in his garden.

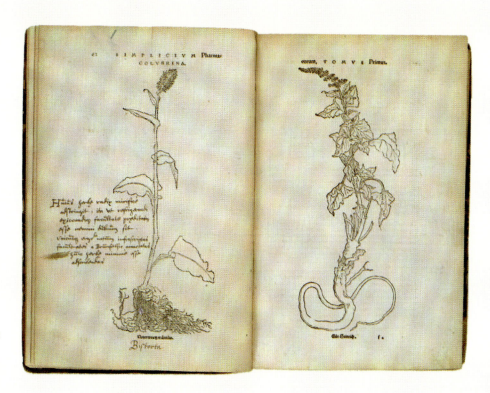

Drawn from live specimens directly onto woodblocks, these woodcuts by Hans Weiditz marked a significant innovation in botanical book illustration.

This does not mean, though, that close observation, analysis, and attempts to group and classify plants did not proceed apace. In 1544 Luca Ghini, 1490–1566, the great Italian botanist and founder of the botanic garden at the University of Pisa, invented the herbarium, a collection of pressed dried plants labelled and systematically classified. This method of display, unlike temporary field observation, enabled the study of a plant form and structure over an indefinite period of time. The co-evolution of the herbarium and the botanic garden continues, with many herbaria serving as important complements to living plant collections. For example, the New York Botanical Garden's continually growing 7.2 million specimen herbarium is used by plant scientists every day.

The great voyages of exploration of this era expanded the collections of living plant material and transformed botanic gardens from places primarily useful to apothecaries and physicians to vital, active centres for the study of ornamental and crop plants. For instance, the collection and later hybridisation of the tulip, a much-prized, purely ornamental garden flower originally found in the wild in central Asia, was due to the efforts of Carolus Clusius (Charles de l'Escluse), 1526–1609, the first prefect of the Hortus Botanicus Leiden. The European diet also gained new foods such as the potato and corn. The gathering and shipment of plant seeds and cuttings from the Americas, Africa, the Middle East, and Asia, as well as from the East and West Indies, necessitated expanded and re-organised layouts of botanic gardens. In addition, because the plethora of new plants being introduced into Europe included many non-medicinal species, exploration and subsequent colonisation required the consideration of botanical science as a discipline in its own right.

The Eighteenth Century Botanic Garden

In order to communicate across vernacular language barriers, medical men and botanists needed a universal classification system that would provide a uniform designation for every plant. The Latin binomial or two-name system—one for genus, the other for species—was the great contribution of the Swedish botanist Carolus Linnaeus, 1707–1778, to natural science. As vernacular languages began to replace Latin as the European lingua franca in letters and published texts, Latin still retained its status as a living language through botany. To this day each new plant, fungus, or other biological discovery is given a Latin binomial.

Portrait of Carolus Linnaeus, 1707–1778, eminent Swedish botanist and inventor of the modern binomial Latin system for plant taxonomy.

While apothecaries wishing to expand their store of herbal knowledge continued to make regional trips in search of medicinal herbs, explorers fostered the development of economic botany farther afield. British landowners in the West Indies found wealth in sugarcane, tobacco, and other cash crops through the exploitation of slave labour. In Mexico and Guatemala, Spanish conquerors forced native people to work producing plant-related dyes, notably indigo and also cochineal, the latter derived from a parasitic insect (*Dactylopius coccus*) harboured on cacti of the *Opuntia* family (the source, incidentally, of the red dye once used for the uniforms of the British army).

The exchange of economic plants from colony to colony became common. After the Dutch discovered coffee in India at the beginning of the eighteenth century, they established coffee plantations in Java, Sumatra, and Bali, their colonies in the East Indies. From Java they shipped seeds to Amsterdam for conservatory propagation in order to produce seeds that were then sent to other conservatories throughout Europe. From this source the French, in 1715, were able to take coffee seeds to start plantations on Martinique in the West Indies. The Portuguese brought seeds from their colony in Goa to Brazil, while the Spaniards brought seeds to Brazil from Cuba.

The Botanic Garden in the Nineteenth Century

Portrait of David Douglas, 1799–1834, explorer and contributor to the Royal Botanic Gardens, Edinburgh.

By the nineteenth century European botanic gardens, most notably the Royal Botanic Gardens at Kew, were sending botanists on plant-hunting expeditions and establishing colonial botanic gardens as outposts to hold and propagate plants destined to be sent back to parent institutions. Scotland's Royal Botanic Garden in Edinburgh was also active in funding expeditions to remote areas. Its roster of intrepid botanical explorers includes David Douglas, 1799–1834, for whom the Douglas Fir (*Pseudotsuga menziesii*) of northwestern America is named, and Robert Fortune, 1812–1880, whose plant-hunting skills are immortalised in the *Euonymus fortunei*. Kew sent Joseph Hooker, 1817–1911, EH 'Chinese' Wilson, 1876–1930, and many other notable botanists to far-away lands, while the Royal Horticultural Society sponsored several plant hunters, including William Forsyth, 1737–1804, one of whose discoveries is honoured by the name of the shrub Forsythia. The Cambridge University Botanic Garden was the beneficiary of numerous herbarium specimens that Charles Darwin, 1809–1882, collected during his five year voyage on the *HMS Beagle* and later sent to its director, John Stevens Henslow, 1796–1861, his former professor and mentor. Partly because of the exciting discoveries of these explorers, botanic gardens also became horticultural showcases, thus stimulating the growth of the nursery industry and the introduction of exotic plants into private gardens during this period.

Eighteenth century view of the Palm House and Waterlily House at Kew, seen with the Campanile rising above the trees.

Simultaneous advances in the manufacture of iron and glass introduced re-fabricated parts into building technology, enabling the construction of large-scale conservatories with curving sides and glass roofs admitting a maximum amount of sunlight, which allowed gardeners to house and protect tropical plants in northern latitudes. In 1836 at Chatsworth, the seat of the Dukes of Devonshire, head gardener Joseph Paxton, 1803–1865, used these malleable materials to pioneer the construction of a 67 foot high conservatory measuring 227 by 123 feet. Called 'The Great Stove', it became the model for his design of the Crystal Palace for London's Great Exhibition of 1851. In 1844 Decimus Burton, 1800–1881, working with iron founder Richard Turner, built the Palm House at Kew. Its dimensions are 363 feet long by 100 feet wide by 66 feet high. Such architecturally striking structures

The Wardian case, a tightly-sealed, glazed plant case, was introduced in the 1830s to transport plants by sea with a greater potential for their survival.

soon became the conspicuous centerpieces of many botanic gardens and parks, particularly in America where grand conservatories were built in Golden Gate Park in San Francisco, Garfield Park in Chicago, the United States Botanic Garden in Washington, DC, the New York Botanical Garden in the Bronx, New York, and the Missouri Botanical Garden in St Louis.

The popularity of the nineteenth century botanic garden coincided with the growth of the public parks movement, and as botanic gardens became places of recreational resort, as well as learning institutions, their collections began to be arranged and displayed within redesigned grounds of a picturesque park-like nature. At the same time, parks became more like botanic gardens with the planting of exotic trees and the addition of display beds for flowers.

The Botanic Garden in the Twentieth and Twenty-first Centuries

The role of the botanic garden as a place to study the medicinal properties of plants persists in a world where approximately 80 per cent of the population still uses herbal remedies. Ethnobotany has become an important branch of botanical studies, wedding sociology with plant science. The recently established Jardin Historico Ethnobotanico at the former convent of Santo Domingo in Oaxaca has as its mission the study of the craft, culinary, and medicinal uses of wild plants by indigenous Mexicans, the conservation of human and plant communities, and the protection of succulents in areas where they are being plundered for sale to commercial nurseries.

Today the global conservation of endangered plants and the ecological niches in which they grow has become an important component of the operations of many botanic gardens. Field scientists work in rainforests and other areas where plants and ecosystems are being destroyed by clear-cutting. The worldwide destruction of native plant species and plant communities has propelled many botanic gardens to undertake educational programmes that stress the role of plants as the primary biological unit upon which all life depends. And some botanic gardens are actually propagating endangered plants and re-establishing them in their natural habitats.

John Stevens Henslow, 1796–1861, founder of the Cambridge University Botanic Garden.

ARGENTINA
CARLOS THAYS BOTANIC GARDEN, BUENOS AIRES

Bordered on three sides by busy streets, the Carlos Thays Botanic Garden is a surprising piece of green in the bustling city of Buenos Aires. The Garden has grown to encompass three distinct landscape gardening styles, hundreds of native and rare plants and an important collection of sculptures. The result of French landscape architect, Carlos Thays, life-long effort, the Gardens were opened on 7 September 1898. Thays had moved from France to Argentina in 1889 where he had been a student of the famous French landscape architect Edouard André. As Director of Parks and Walkways for the City of Buenos Aires, Thays requested 17.3 acres of land in 1892, with the aim of developing parkland in the Palermo area. Thays chose this particular neighbourhood because it was close to the main city parks, the Bosques de Palermo, the Japanese Gardens of Buenos Aires and the Buenos Aires Zoo. His idea was to make "a garden of adaptation", a place where the people of Buenos Aires could spend their recreation time but also a site for botanic research and education for the Argentine people as a national resource.

Thays and his family took up residence in the English-style mansion situated on the grounds, from where he spent the next six years meticulously designing the entirety of the Garden's landscape. The mansion remains the main building on the grounds, and a monumental reminder of the rich history concealed within the Garden's charming landscape. The five winter houses in the Garden are some of the most beautiful and unique in the region, indeed the largest was brought across from France in 1900. Built in the Art Nouveau style, it was exhibited in the 1899 Paris Exhibition, and is thought to be the only greenhouse of this style still in use today.

The Garden was designed in, and still maintains, various gardening styles; the symmetric, the mixed and the picturesque. The mixed Roman gardens hold species of plants that the first century Roman writer, Pliny the Younger, describes as having at his villa. The French Garden displays the French symmetric style of the seventeenth and eighteenth centuries, and an Oriental Garden features the picturesque beds characteristic of eastern Asia.

In other areas of the Botanic Garden, plants are ordered to re-create environments from across the globe. The Asian area shows typical Oriental varieties such as the Ginkgo Bilobas plant and Africa is represented by brackens, palms and gomeros. The Oceania grounds display species typical of the region such as acacias, eucalyptus and casuarinas, and from Europe there are oaks, hazelnut trees and olmos plants. The specimens are systematically ordered by taxonomic qualification, and contextualised in settings that highlight their beauty and unique qualities, making the Carlos Thays Botanic Garden a delight for both the serious botanic enthusiast, and the recreational visitor.

Though the Garden displays samples of flora from all over the world, it specialises in Argentine plants and trees. Thays spent significant amounts of time researching the forest characteristics

of Argentina, replicating important and endangered systems within the grounds. Winding paths lead visitors across the various beds and themed areas to a brook on the farthest limits of the park. An impressive collection of over 30 works of art are scattered throughout the grounds, including sculptures, busts and monuments acquired throughout the Garden's long history.

The Garden also houses some important academic institutions, such as the Botanical Museum and Botanic Library, a collection of about 1,000 books and 10,000 periodicals accessible to all visitors, culminating in a valuable botanic resource for scholars and hobbyists alike. The main brick building also houses the Municipal Gardening School and the herbarium for more serious horticultural study.

tag: White and yellow Waterlily (*Nymphaea*). opposite: Interior view of the main greenhouse at Carlos Thays Botanic Garden. above: *The Spring,* by Lucio Correa Morales. This sculpture can be found near one of the Garden's entrances.

ARGENTINA
BOTANIC GARDENS OF THE NATIONAL UNIVERSITY OF BOTANY, CÓRDOBA

Located in the Quebrada las Rosas district of the Argentine city of Córdoba, the Botanic Garden was established in 1997, created in the interest of furthering education and the conservation of natural resources. The initial site was extended in 2005 to give the Garden a total area of only 12 acres. However, the limited space has been utilised sensitively and effectively to display a selection of subtropical and native vegetation.

The collections of trees and plants at the Botanic Garden have been devised in terms of naturalistic, easy viewing. In order to make learning a pleasurable process, several thematic promenades or 'strolls' have been devised such that groups of species can be leisurely appreciated in a contextualised setting. For example, the Stroll of Flowers, where the visitor passes through the vivid flora and heady aroma of subtropical blooms, the Stroll of Palms, with an abundance of architectural foliage and the Stroll of Pebbles, which displays four zones of rockery and mountainous plants. Walking through the simplified thematic scheme is a dynamic and accesible way to view the maximum range of taxa without feeling overwhelmed. A particularly interesting and commendable display is the Stroll of Urban Trees, where these often unsung saviours of the urban environment are glorified, encouraging a recognition and

greater involvement with common species that are often ignored. These include the pyramidally structured cypress, poplar and eucalyptus, as well as many other native perennial shrubs.

Finally, two 'Strolls' have been devised to showcase economic and utilitarian plant crops; the Stroll of Aromatic and Medicinal plants is an olfactory delight, where three central areas of herbs are flanked by shrubs used in medicine or gastronomy. The Pasture Stroll completes the set, exhibiting domestic and industrial grasses and cereal crops, as well as the grasses native to the Argentinean Pampas lands. Ornamental attributes of exotic grasses are highlighted with structured, grouped planting.

The Infiernillo Stream runs through the Garden, creating several picturesque viewpoints from which to admire the exhibits; a pergola situated near the stream is planted with perennials and

other summer and winter flowering vines and plants, such that it is always naturally adorned. Many of the plants such as jasmine and Santa Rita are frequently used in the gardens of the city; ornamental planting of local plants is used to promote horticulture and display innovative planting ideas.

A glasshouse allows for the cultivation and protection of native and non-native species, and the Garden hopes to add a herbarium in the near future to house research material. Other visitor-friendly features include a labyrinth, a terraced bar facing onto a lagoon and a play area for children, which utilises trunks and stumps to introduce contact with the botanical world.

tag: Cactus (*Gymnocalicium*). opposite: View of the Glasshouse. above: View of the Garden from inside the main building. all photographs: Eugenia Viviana Alvarez/Botanic Garden of Córdoba.

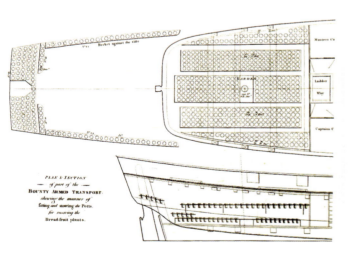

Wrapped around Farm Cove on the shores of Sydney Harbour, directly east of the Sydney Opera House and Circular Quay and bordering the west side of the Central Business District, the Royal Botanic Gardens in Sydney have one of the most spectacular and stunning settings of any botanic garden in the world. The Botanic Gardens were founded in 1816 by Governor Lachlan Macquarie, on the site of what had been the first farm on the Australian continent (established in 1788) known as Farm Cove. In 1817, Charles Fraser was appointed as the First Colonial Botanist, thus beginning a history of collection and the study of plants that persists today, making it the oldest scientific institution in Australia. The open parklands of the Domain surround the Gardens, which in colonial times was the Governor's buffer of privacy between his residence and the penal colony. Roads and paths were constructed through the Domain in 1831, opening up the region for people to enjoy as a recreation area, complimented by the breathtaking vistas of the Botanic Gardens.

The three million visitors who pass through the park each year are able to visit the 17 themed gardens designed to showcase the diversity of Sydney's collection. The formal garden display of the Government House Grounds, with its manicured lawns and large scale shrub planting, is a regal reminder of the Garden's colonial history. The Rare and Threatened Plants Garden features plants from around the world that are rare or on the brink of extinction. The Tropical Centre offers a chance to discover plants from tropical ecosystems, including bat flowers, jade plants and the colourful heliconia. The Cadi Jam Ora: First Encounters Garden features the original plants that grew on the site before its development as a botanic garden while signposts and exhibitions tell the story of the Cadigal people, the traditional Aboriginal owners of the Sydney city area.

From its inception, Sydney's Botanic Gardens have played a major role in the acclimatisation of plants. The over 70 acres of land are home to more than 45,000 plants and 10,000 plant groups. The National Herbarium of New South Wales, which is situated in the Robert Brown Building on the Woolloomooloo side of the Gardens,

is home to over one million preserved plant specimens, the earliest of which was collected by Joseph Banks and Daniel Solander from Botany Bay in 1770. The Herbarium is a major research centre for Australian plant studies, providing reference material for studies of Australian native plants and their relationship and classification. Scientists conduct research in plant systematics, including classification studies, DNA fingerprinting and the ecology of New South Wales. A public reference collection for native plant identifications are also available at the Herbarium through the Botanical Information Service. The Botanical Library, also located in the Gardens, is home to one of the most extensive collections of botanical, horticultural and ecological books, magazines, videos, photographs and publications, as well as material on related historical and scientific areas of research.

The mission of the Gardens is to inspire the appreciation and conservation of Australian flora, indeed some of Australia's oldest and most important vegetation is located in the Gardens, including remnants of native trees that preceded English settlement such as the Sydney Red Gum (*Angophora costata*) and the three Forest Red Gums (*Eucalyptus tereticornis*). Australia's oldest planted trees

are also situated here, the Hoop Pine and the Giant Watergum, both of which were planted between 1820 and 1828 in the Palm Grove. Apart from the impressive collection of historic flora, the Garden also plays home to a number of native animal species including the Sulphur-crested Cockatoos, White Ibises, Brush-tailed Possums and Flying Foxes and the sole colony of Pearl White Butterflies (*Elodina angulipennis*) in the Sydney Basin lives in the Rare and Threatened Garden.

With dual aims of propagation and conservation, the Royal Botanic Gardens in Sydney offer abundant resources for botanic study, while offering a fascinating collection of native Australian flora whose origins and evolution track the very history of the country itself.

tag: *Clivias* in the Botanic Gardens' Palm Grove. previous pages top left: Map showing the water depth in Botany Bay. previous pages bottom left: Sketch of the potting areas aboard the *HMS Bounty*. previous pages right: The Sydney Tropical Centre, which is home to a variety of mountainous and woodland plants. photograph: Jaime Plaza. above: View of Sydney Opera House and Harbour Bridge viewed from the Royal Botanic Gardens. photograph: Jaime Plaza. opposite: Glass sculptures, made by Dale Chihuly for the Sydney Biennale.

Set out in the picturesque landscape style, the Royal Botanic Gardens, Melbourne, is among the prettiest botanic gardens in the world, with its sweeping lawns, winding paths and stunning lakes, which showcase a series of magnificent vistas presenting a revelation at every corner. Established in 1846, on what was a swampy, rocky area on the South bank of the Yarra River, the area began its transformation into a botanic garden in 1846 so that today, they are home to over 12,000 species and 51,000 individual plants, many of which are endangered in the wild, or even extinct. The mild Melbourne climate allows for an interesting mix of tropical and temperate plants, and a long history of plant collection by the Gardens, means that they have a large range of vegetation, both native to Australia and from across the world. Black Swans, Bell Birds, cockatoos and kookaburras are just some of the native wildlife that make their home in the Gardens.

A number of different features and institutions are situated on the 94 acres of gardens at Melbourne. The oldest is the Herbarium, founded in 1853 by the then director, Ferdinand von Mueller, which houses over 1.2 million pressed plant species including native Australian specimens and examples from across the globe. The collection of nineteenth century specimens is especially impressive, mostly due to von Mueller, who was an avid plant collector and amassed an extraordinary range of plants from every corner of the globe. The collection includes examples from colonial Australian history, including specimens collected on Cook's voyages to Australia and Burke and Wills first forays into the outback.

The Gardens are set out in a number of groupings, or collections, chosen for their value, rarity, diversity and interest. The main groups are the Australian Rainforest Walk, Cacti and Succulents, the Californian Collection, Cycads, Eucalypts and Fern Gully collections, Grey Garden, the Herb Garden, Long Island (a newly-developed collection of Australia flora), Oaks, the New Zealand Collection, Perennial Borders, a Tropical Displays Glasshouse, Viburnum Collection and the Water Conservation Garden. The latter is one of the most important of the Gardens' initiatives, as in the arid Australian landscape, water availability and usage is a vital ecological issue.

The Royal Botanic Gardens, Melbourne, place a strong emphasis on scientific investigation into taxonomy, systematics (the study of relationships between organisms) and ecology within Victoria. Since its early days it has been involved in plant research and identification. They are a key scientific institution for furthering knowledge and understanding of plant biodiversity and its greater impact on surrounding environments as well as the conservation of Australia's plant biodiversity. They are also interested in passing practical information on to the local community for use in their own gardens, providing a large education service, free and accessible information about good-gardening practice, and advice on water conservation.

The many and varied events and exhibitions that are held at the Gardens provide good initiatives to get people of all ages to visit the grounds and participate in Melbourne's rich botanical programme. Events such as the Rockpool Twilight Evenings, where the Gardens are open late into the night, bring in a new demographic, and the range of planting beds, including the Ian Potter Children's Garden, help to raise the Garden's profile as a leisure destination. Other features of the Garden, include one of Australia's most comprehensive botanic libraries, a scientific centre, a garden shop, an ornamental lake and a Children's Garden, all of which makes visiting a perfect outing for locals and tourists alike.

tag and opposite bottom left: The Perennial Border at the Royal Botanic Gardens Melbourne which features large drifts of perennials interspersed with sculptural, ornamental plants. opposite top left: View from Taxodium Bridge across the Ornamental Lake. opposite top right: View of the Rose Pavilion. opposite bottom left: Moreton Bay Fig (*Ficus macrophylla*). all photographs: Janusz Molinksi.

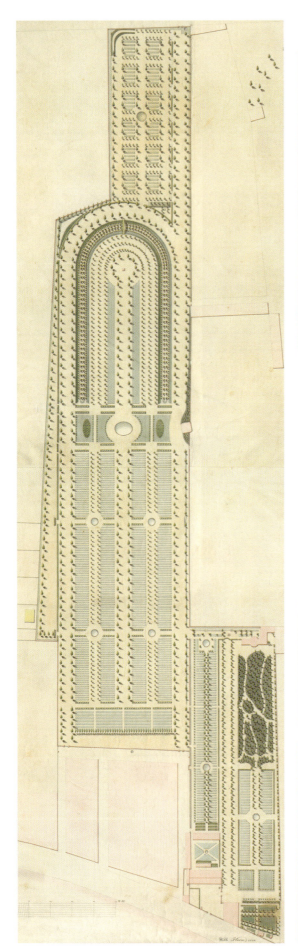

AUSTRIA

VIENNA UNIVERSITY BOTANICAL GARDEN

With regal beginnings, the Vienna University Botanical Garden was founded by Archduchess Maria Theresa of Austria in 1752 and has continued to play home to some of botany's most revered historical figures. Nikolaus von Jacquin was one of the Garden's first directors, who not only taught botany, but also chemistry and other related disciplines. His son, Joseph von Jacquin, succeeded him as director, as did a number of other leading botanists in turn, including Stefan Endlicher, Eduard Fenzl, Anton Kerner von Marilaun, Richard von Wettstein and Fritz Knoll, all of whom were directors of the Botanical Garden as well as professors of botany in nearby universities.

Still situated in its original location, the Vienna University Botanical Garden forms the basis for today's botanical garden, associated with the Institute of Botany at the Faculty of Natural Sciences at the University of Vienna. After several changes, the Botanical Garden now boasts a unique fusion between a nineteenth century landscape garden and a modern display of plants contained within state of the art scientific research beds.

Under the instruction of Richard von Wettstein, the first genuine institute building was erected and opened in 1905, allowing for a cutting-edge treatment of systematic botany that included the burgeoning sciences of mycology and embryology. The building provided ample space for a library, a herbarium and a botanical museum, but lacked space for botanical instruction. The Garden therefore sought to renovate the structure, with works concluding in 1992, and the much-modified building now provides a broader range of facilities for students and staff. Lectures, courses and research facilities are offered in botany, cytology and genetics, microbiology, ecology, and pharmacy and several hundred students receive training in nearly 200 basic and advanced courses every year.

Von Wettstein was succeeded by Fritz Knoll as director of the Garden, who—together with Karl von Frisch— founded experimental flower ecology. At the end of the Second World War, the institute, all the greenhouses, and the entire Garden area were ravaged by enemy bombs, and the Garden needed to be entirely rebuilt if it was to survive at all. In 1947, the grounds opened after a sabbatical of two years with Lothar Geitler—highly respected as a cytologist and algologist—as its head, and the mandate to restore the grounds to their former glory. The Garden now covers nearly 20 acres of land, and contains more than 9,000 species of plants, including cultivated woody tropical plants, notably of such families as *Annonaceae*, *Rubiaceae*, *Gesneriaceae*, *Bromeliaceae* or *Orchidaceae*. A number of smaller areas in the Garden are devoted to specific themes, such as useful plants, succulents, Alpine floras, Austrian vegetation, plant genetics and evolutionary plants. The Garden has also acquired a great variety of rare plant species, including live fossils or insect-eating plants

The greenhouses were originally built between 1890 and 1893 but suffered severe damage during both world wars, and have now been renovated or rebuilt. Due to a lack of space, most of the greenhouses are used for scientific purposes only and not open to the public. However, a permanent rainforest exhibition for visitors has been established in one greenhouse and in the warm climate of the Tropenhaus (Tropical House), there is an impressive collection of cacti and succulents, accompanied by a varied planting of alpine plants. Though visitors cannot access the entirety of the Vienna University Botanical Garden, it remains a forerunner of botanic innovation and a charming setting for academic research.

tag: A Pasque Flower plant (*Pulsatilla grandis*) in full bloom. photograph: Rudolf Hromniak/Vienna University Botanical Garden. opposite left: Historical map of the Vienna University Botanical Garden, 1832, by Wilh. Horn. photograph: Archive and Rudolf Hromniak/ Vienna University Botanical Garden. opposite top right: An Orchid (*Bulbophyllum lakatoense*) from Madagascar. This plant is one of the smallest in the Vienna University Botanical Garden's collection. photograph: Martin Summhammer, Rudolf Hromniak and Frank Schumacher/Vienna University Botanical Garden. opposite bottom right: View of the Main Alley in the Vienna University Botanical Garden. photograph: Rudolf Hromniak/Vienna University Botanical Garden.

NATIONAL BOTANIC GARDEN OF BELGIUM, MEISE

Just a few kilometres away from the world famous Atomium lies the National Botanic Garden of Belgium. Located in Meise, this establishment contrasts with its iconic cousin; steeped in noble tradition and an appreciation of nature, the history of one balances the scientific ideals of the other.

The Garden is housed in the historic Bouchout Estate, surrounding Bouchout Castle. The earliest traces of the noble Bouchout dynasty can be found in the twelfth century; the most famous inhabitant of this estate was Charlotte, Empress of Mexico and sister of King Leopold II. After she died, the Belgian Government bought the estate to house the National Botanic Garden. Its 200 acres comprise wild woodland, several gardens and historically landscaped areas, an orangery with a walled garden, nineteenth century follies and a public glasshouse complex covering more than two and a half acres.

Under the reign of the Austrian Emperor Joseph II, a botanic garden was established in Brussels at the turn of the eighteenth century. Following this, the crown societies for horticulture and botany underwent several operational changes over the nineteenth century, eventually settling on the current state-run incarnation at the present site in Meise. From 1939 onwards, the Garden started moving into the Bouchout estate; the Second World War slowed down the process also development persisted apace and in 1967 the institution was officially renamed the Nationale Plantentuin van België or Jardin Botanique National de Belgique (National Botanic Garden of Belgium).

The major mission of the modern Garden became scientific research in botany and horticulture, but specifically and somewhat unusually, Central Africa lies at the core of its research and collections. The Garden cultivates many African plants in its living collections; the Central African Herbarium is arguably the largest in the world and the garden is actively involved in several African projects, particularly in the Congo. The coffee family collection or *Rubiaceae* is another focal point of the Gardens' research. Largely indebted to former global trade links, this economic crop family

has been studied here for more than a century and Belgium was instrumental in describing the economically important Robusta coffee (*Coffea canephora*) for the benefit of botanists worldwide. Presently the Garden holds one of the largest living collections of *Rubiaceae*, which gives it a particular point of interest to visitors who will recognise the value of this everyday plant.

In total, the Garden cultivates approximately 18,000 different types of plants from all over the world. The more delicate species are housed in an impressive array of more than 60 different glasshouses, the largest of which are accessible to the public. The Mediterranean House, the Victoria House and several Rainforest Houses containing creepers, climbers and various tropical

blooms are but a few examples of the collections housed in these impressive structures. Two of the glasshouses have a central rather than a systematic theme: the Mabundu House holds tropical economic plants, while the Evolution House traces the 500 million year history of the plant kingdom. The elegant Balat House, dating from 1853, is interesting both for its content and for the collection of flora it houses, where a rare agave collection resides. The Balat House lies at the centre of the Herbetum, a systematic and formal garden presenting herbaceous plants. A collection of flowering shrubs and trees, known as *Fruticetum*, surrounds the Herbetum and offers year-round interest. Further into the Garden, mixed borders break up the formal and ornamental sections; to satisfy dendrologists, the park maintains an oak arboretum, maples and a conifer collection as well as a lovely rhododendron wood which becomes a veritable fairyland when in bloom. The Orangery, dating from 1818, is set in a picturesque landscape, featuring a large pond that boasts a particularly stunning autumn bloom. In front of the Orangery, south-facing terraces provide more susceptible

plants refuge from the cold. The historic Walled Garden behind the Orangery also offers shelter to semi-hardy plants and a peony garden is the latest addition to the living collections, whilst a medicinal garden, a hydrangea display, and many more specifically landscaped sections also contribute to this impressive selection of flora from around the world.

The general exhibition of plant life together with the architecture in the park merit an enjoyable and interesting visit; the incongruous and specially focused research and exhibitions on Central African vegetation mark the National Botanic Garden of Belgium as a distinctive and expert institution unique to Europe.

tag: Pride of Madeira (*Echium candicans*). opposite top: The Myriad Glasshouses and boiler chimneys of the Van Houtte Nursery. opposite bottom: The Balat House in the centre of the Systematic Garden. © Nationale Plantentuin, Belgium. above: A Waterlily (*Victoria amazonica*), in one of the National Botanic Garden's ponds.

BELIZE
BELIZE BOTANIC GARDENS, SAN IGNACIO

The Belize Botanic Gardens began when husband and wife team Judy and Ken duPlooy (originally from Maryland and Zimbabwe, respectively) ignored advice from family and friends and moved from their comfortable home in Charleston, South Carolina, to the rainforest of Belize in 1988. The duPlooys travelled with their five daughters, aged four to 16, from the United States through Mexico and into the Cayo district of Belize. When the family arrived, they decided to settle on a large cleared farm, and soon began building a home and small bungalows on the site. Ken, then a novice, soon began planting and landscaping the plot, and was encouraged by the quick growth rate to become an amateur botanist, planting an increasingly large variety of flora on the grounds.

In 1994, the Gardens expanded further when the family bought a property adjacent to the Macal River and regenerated a crop of native tropical plants and fruit trees. This particularly diverse collection captured the attention of scientists, plant enthusiasts and farmers from across the globe. In 1997, the Gardens earned official recognition and were registered as the Belize Botanic Gardens. Now 45 acres strong, the Gardens boast a distinctive collection of native and exotic plants situated in the valley of the Macal River and surrounded by the Maya Mountain foothills. Although Ken died in 2001, the Gardens remain a dedicated family affair with a small and dedicated staff of eight who work to encourage sustainable agriculture, maintaining conservation collections and engaging visitors in conservation education.

The Gardens focuse primarily on the local flora of Belize but also displays exotic plants from tropical regions around the world. Belizean plants, as well as economically, botanically or horticultural important species such as palms, cycads, bromeliads, passion flowers and hardwoods pepper the gardens, culminating in an enchanting experience for visitors and scientists alike.

The Gardens have an orchid house with over 300 varieties of native Belizeian orchids protected from the constant danger they face due to development and logging. Reaching back to the native cultures of Belize, the Plants of the Maya Trail is a walking expedition through the history of the Mayan people and their innovative use of rainforest plants. Although Belize is now a mix of Spanish, English, Mestizo, Garifuna, Creole and Mennonite culture, it is thought that nearly one million Mayans historically inhabited Belize, and the trail shows visitors how the Mayans used plants to create clothes, tools, buildings, and medicines.

In the institution's characteristically creative style, staff at the Belize Botanic Gardens have recently created Bauble, an open access software programme that helps manage live plant collections and is available free to botanic gardens, herbaria and arboreta worldwide in a bid to make collection records available online and integrate plant records with the Global Biodiversity Information Facility. The Gardens also work to promote sustainable agricultural growth and the healthy harvest of local palm plants, with a collection of 40 native palm species as well as 50 such species from around the world, including two Cuban palm species that are currently endangered. Succulent tropical fruits such as lychee, rambutan, mangostten, bredfruit, jackfruit, mango, avocado, starfruit, jaboticaba, Mammy Apple and Malley Apple bloom throughout the year, providing a fragrant and beautiful setting for visitors throughout the summer and winter months.

The Gardens work with local farmers to introduce sustainable crops and the use of organic pest controls and fertilizers, as conventional farming methods in the area are destructive to local plant life. Workshops provide valuable information to farmers on the profitability of sustainable farming where techniques such as agroforestry—the use of trees to help improve farming methods—is promoted. The Gardens also run extensive education programmes for visitors and schoolchildren that explain the important role of plants in human health and the environment. From humble beginnings as a novice garden, the Belize Botanic Gardens has grown to be contribute significantly to the international botanic community.

tag: View from the Macal River trail at Belize Botanic Gardens. photograph: Brett Adams. opposite: Aerial view of the Belize Botanic Gardens. photograph: Mike Green.

MONTREAL BOTANICAL GARDEN

Brother Marie-Victorin was the tireless founder and ally of what was to eventually become the Montreal Botanical Garden. President of the city's Biology Society, he is on record as having suggested that Montreal build its own botanic gardens as early as 1919, although it was not until 1929 that Montreal mayor, Camillien Houde, authorised the development of the Garden. Marie-Victorin was dedicated to the Montreal Botanical Garden until his death in 1944 and promoted it at every opportunity, leading specimen collection expeditions and protecting it from being converted into a military flight school during the Second World War.

Oddly, the Garden was founded at the height of the Great Depression, when Montreal was hit hard by homelessness and unemployment, and almost all construction and urban development was halted. It seems amazing that such a celebratory Garden was created during so somber an era, yet perhaps reinforces the faith architects and urban planners have in the social consequences of such gardens. Although the primary function of the Montreal Botanical Garden was reportedly to educate the public and students of horticulture, one must not underestimate the tranquilising, sometimes uplifting, effect of a large, natural space amidst the mess of collapsing industrialisation.

The site now stretches to over 183 acres and is distinguished from other gardens by several unique features including an insectarium, which chronicles the role that insects play in our biosphere, a wooden tree house, which is an interactive museum devoted to the importance of trees, and the new Garden of Innovations that showcases the latest plant varieties and trends in landscape design. Geographically themed gardens include the Japanese and Chinese Gardens (the latter the largest of its kind outside of China) and, on a more a vernacular note, the Alpine and First Nations Gardens. Canadian maple, birch and pine trees populate the First Nations Garden, and numerous totem poles and exhibits demonstrating traditional First Nations artwork pepper the serene landscape. The Alpine Garden has several paths winding over a rocky outcrop that is covered with tiny, delicate alpine plants. Both gardens seek to draw attention to the distinctive northerly position of the Montreal Botanical Garden, and how its natural landscape can be both hostile and sympathetic towards its native flora.

Few botanic gardens are situated in such inconsiderate climates, Montreal spends almost half of each year blanketed by snow, and much of the Botanical Garden becomes inert. Thankfully (and deliberately) the winter months are when the Garden's

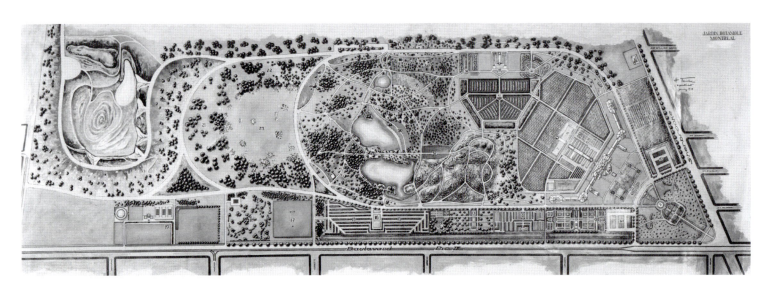

indoor greenhouse complex is at its most splendid. Large, colourful spectacles are annually arranged to contrast with the harsh outdoor climate. There are lush greenhouses populated with tropical rainforest plants—blooming orchids, aroids and gesneriads—and arid greenhouses containing African and American desert plants. The Hacienda Garden features pools, terraces and indoor courtyards to evoke the atmosphere of Hispanic gardens, and houses a remarkable collection of perennial plants from the Mediterranean region. Among the species on display are the captivating Lithops—also known as the 'living stones' because of their ingenious camouflage, designed to protect them from grazing herbivores in the wild. Mainly a recreational garden, the variety and quality of exhibitions available at the Montreal Botanical Garden makes it one of the most important and beautiful gardens outside of Europe.

tag: A spring flower in the Garden. © Jardin botanique de Montreal opposite: Historical plan of Montreal Botanical Garden. © Jardin botanique de Montreal. above: Main Entrance of the Montréal Botanical Garden. photograph: Michel Tremblay. © Jardin botanique de Montreal. below: View of the Garden's lake.

UNIVERSITY OF BRITISH COLUMBIA BOTANICAL GARDEN, VANCOUVER

On the tip of Point Grey in Vancouver's University of British Columbia (UBC) lies Canada's oldest established university botanical garden. UBC Botanical Garden has evolved over its 90-year history into a diverse and expansive series of several microclimates, duplicated habitats and themed gardens. While the Alpine Garden offers the opportunity to ramble through from the Andes to Canary Islands and Asia Minor, the Asian Garden—the largest of the group—comprises flora and fauna gathered from Tibet, Japan, China, and Manchuria.

The Botanical Garden began life at Essondale, just East of Vancouver, in 1912. The government set aside two acres of land for the study of plants from all over British Columbia. The region's first provincial botanist, John Davidson, was appointed to complete this endeavour, travelling extensively around the area to collect plants for the Garden and Herbarium. Four years later Davidson had amassed some 9,000 species of plant, and the Garden was relocated—with a view to its expansion—to Point Grey, where a new university campus had been established. Many of the plant species collected by Davidson can be found in the Native Garden, which is reserved for showcasing the vegetation that makes up the habitats of Vancouver, as well as the insects, birds and frogs that dwell within typical west coast settings.

The diversity of plant-life here and the rigour with which it has been formed into a number of unique bedding systems is illustrative of UBC's world-renowned commitment to research, conservation and teaching. Fostering strong links with the horticultural industry, the Garden established its Plant Introduction Scheme of the Botanical Garden (PISBG), enabling the propagation of endangered flora and innovating new species of plant life. With all its diversity of exotic forms, significant areas of the Garden are devoted to a horticulture that is entirely Canadian, with a dark, silent lake at the centre, and its peripheries marked by towering pines.

The Food Garden, a favourite amongst children, and growing ever more popular amidst the current interest in organic food, contains experimental breeds of fruit trees, vegetables and berries and distributes its produce amongst local soup kitchens at the end of the harvest season.

Although widely-renowned as a research centre, the Garden is less well-known and therefore less visited than most tourist attractions in Vancouver, so the visitor can still expect to wander through the Garden in relative peace, contemplation or study. Other wonders of the grounds include a Japanese garden, the Nitobe Garden, and a physic garden, mapping an extensive history of medicinal plant-use. A Carolinan Forest, providing a rare example of this kind of hardwood landscape, will be the next addition to the Garden, featuring plants native to the west coast of North America.

tag: *Schima sericans*. opposite top left: The David C Lam Asian Garden. opposite top right: The Nitobe Memorial Garden. opposite bottom: UBC Botanical Garden entrance in winter. below: The Physic Garden. all photographs: Daniel Mosquin/UBC Botanical Garden.

UNIVERSITY OF CHILE ARBORETUM, VALDIVIA

With more than 500 species of native and exotic trees, the University of Chile Arboretum harbours the largest sample of tree and shrub species in the country. Located in the Los Lagos region in the southern province of Valdivia, the Arboretum stretches over 135 acres of land maintained by the University of Chile and the Institute of Silviculture, surrounded by natural forest and rainforest growth. The humid and warm climate typical of the area provides for a rich and ample classroom within which to study the lifecycle of temperate forests.

The Arboretum was erected due to the initiative of a group of professors in 1971, who wished to preserve the native natural landscape through an area dedicated to scientific observation and teaching. It was laid out in the northern sector of part of the University Campus, and extends to the edge of the nearby Saval Natural Park. The main objectives of the Arboretum, apart from the facilitation of academic study, were set out as the formation of a natural reserve for endangered arboreal and arbustival species both native and introduced, the production of seeds for conservation and exchange, and the maintenance of a collection of commercial bamboos. Other varieties of tree cultivated in the Arboretum include those of the native Valdiviano rainforest, examples from Central Chilean forests, firs and coniferous varieties introduced from Germany, as well as a comparative display of the native Nothofagus, with variants of the species from New Zealand.

The Arboretum was created with the special interest in preserving rare and endangered species, including many beautiful varieties native to the area. Amongst the endangered species, the Arboretum offers the rare opportunity to view examples of the *Sophora toromiro skottsb*, a species endemic to Easter Island, but which due to historically intensive use by the Rapa Nui for its wood, is sadly extinct in its natural habitat. Visitors can enjoy a rare view of these ancient species, now thriving in but a few specialised arboretums in the world.

Rare species and commercially viable varieties are subjects of particular interest in the Populetum, which is an area of eight and

a half acres that the University devotes to the study of genetically modified variants. The hybrids are created to fulfil different objectives such as resistance to fungi, quality of wood, and adaptability to extreme climactic conditions.

The Arboretum is, however, well equipped for the lay visitor, providing an informative and pleasurable experience to those with no scientific motive. These include fragrant fir and cypress forests, gardens displaying the visually arresting forms of bamboo grasses and a lagoon of heady lotus flowers, providing for a charming setting for a relaxed day out. The public areas of the

Arboretum are intersected by landscaped footpaths, from where the visitor may spot various examples of local wildlife. Indeed the grounds of the Arboretum is a favourite haunt of several species of woodpecker and various other colourful birds and woodland animals native to Chile.

tag: Chilean Firebush (*Embohtrium coccineum*). opposite top: View of the Arboretum Pond surrounded by the Chilean Bamboo Collection. opposite bottom: The Forestry Trees Plant Collection, the biggest collection of its type in Chile. above: A small-scale specialist nursery comprising two polythene houses and a large shaded area for potted plants.

CHILE
NATIONAL BOTANIC GARDENS OF VINA DEL MAR

The National Botanic Gardens of Vina Del Mar began in 1918 as Pascual Baburizza private garden. Now bordering on the city's industrial zone, it comprises nearly 100 acres of woodland, stony paths, picturesque ponds, rose gardens, and rare plants and trees handed over to the public in 1951.

The Gardens' unique features include endemic plants of the Juan Fernandez Archipelago and (in conjunction with the Tomoriro Management Group based at Kew Gardens) cultivates one of the only collections of the Toromiro (*Sophora toromiro*) in the world, a tree native to Easter Island that is now completely extinct in the wild. The Gardens pride themselves on its preservation of protected flora from Central and South America, indeed they cultivates 779 species of plants, 43 per cent of which are threatened. The Gardens operate with a mandate of conservation enacted according to core principles that are based on an understanding of the diversity of the plant kingdom, whose continual existence depends upon a balance within, and between, ecosystems.

The layout of the park belies its location. Despite being situated in a dry and hilly region, the Gardens contain a multitude of different areas that would not ordinarily coexist. We find the French Garden, which is laid out with borders in a formal nineteenth century style, situated alongside the Canopy Zone whereby a circuit of bridges between the tall trees in an area of dense forest enable visitors to view the Gardens from a unique vantage point. There is also an artificial lake that enhances the aesthetic quality of the Gardens while providing the perfect home for the many species of aquatic birds that frequent the area. These different habitats are linked through a series of meandering paths and more formal decking and bridges with guide chains. As for the future, there is currently a large development zone in place, at present consisting mostly of elevated, arid land, the intention for which is to encourage the diversification of the Gardens' burgeoning collection.

tag: The Archipelago Plant Collection of Juan Fernández. below left: View of the Olive Grove. below right: View of the French Garden and Central Paddocks. opposite: The Archipelago Plant Collection of Juan Fernández. all photographs: Patricio Novoa/Jardín Botánico Nacional.

CHINA
BEIJING BOTANIC GARDEN

Beijing Botanic Garden is one of the most forward-thinking gardens active today. The institution's practise is based on principles of conservation, particularly that of native flora from northern China. Alongside this mandate, the Beijing Botanic Garden aims to provide educational, research and recreation facilities to the rapidly growing city of Beijing.

The Garden was founded in 1956 with financial support from the central government. Following a period of rapid development, the Garden's activities were all but halted by the Cultural Revolution during the early 1960s. It was not until after 1990 that renewed interest in the park fostered the initiative of several construction and planting programmes, ultimately raising the Garden to international status, indeed today the Beijing Botanic Garden is thought to be one of the most innovative and important gardens in the world.

The main conservatory on the grounds is among most futuristic to be found within the limits of any botanic garden. It was erected in 1999 and boasts a substantial 6,500 square meters of display space in which a staggering 3,000 taxa of plants are grown. The Conservatory is home to a rainforest section, a cacti and succulent display, an orchid and carnivorous plant collection, an alpine house and many other varieties of planting beds.

True to Beijing's botanical history, the Garden has an impressive collection of bonsai trees in the Penjing (bonsai) Garden. Here penjing from all over China are displayed, with each region's contribution expressing different features and characteristics specific to the different traditions alive in Chinese penjing practise. There are also outside miniature trees that are over 100 years old, including a ginkgo tree that has been growing for an astounding 1,300 years! The Garden provides an elegant and informative surrounding in which to enjoy and observe this ancient Chinese art in all its intricate glory.

The more historically important buildings in the Beijing Botanic Garden include the Sleeping Buddha Temple, which was built during the Tang dynasty, 618–907, and renovated during the Yuan, Ming and Qing dynasties. The temple complex leads the visitor through four decadently appointed halls and courtyards, with a large coloured glass screen overlooking a pool bisected by a stone bridge, and the Shouan Mountain rising behind the picturesque scene. The temple gets its name from the 1,321 bronze statue of the sleeping Buddha, which is contained in one of the Garden's elaborate halls.

The Rose Garden cultivates over 700 varieties of this most cherished flower, spanning over 17 acres. This area is designed in a characteristic European style, with a hidden fountain and cascade providing the perfect backdrop for the hybrid tea roses, miniature roses, antique roses and wild roses that charm visitors to the grounds every year. More representative of Beijing's botanical culture, there is also an Ornamental Peach Garden that has over eight acres bursting with the cheerful blooms and twisting branches of one of China's most important botanic emblems.

tag and right: Potted display outside the Beijing Botanic Garden Temple. previous pages: View of the Iris and fossil centrepiece in the Formal Garden. opposite: Fossil Trees in the Beijing Botanic Garden. Image courtesy of Heather Angel/Natural Visions.

The study of plants within the tradition of Chinese herbalism and botany has been developed over centuries; nowhere is this more evident in the present day than in Kunming Botanical Garden in Yunnan, associated with the prestigious Kunming Institute of Botany. Over the last 66 years the Institute, together with the help of its Botanical Garden, Herbarium and affiliated organisations, has become a vitally important centre for the study of the plant resources of southwest China.

The Garden was first built in 1938 for the purpose of cultivating rare and endangered flowers, medicinal herbs and major trees. Since its inception, the Institute and its gardens have endured several shifts in affiliation and management, being taken over by the Chinese Academy of Sciences in 1951, as the Kunming Station of the Institute of Botany, and wavering under the control of the Yunnan Institute of Botany before being reabsorbed into the Chinese Academy in 1979. Throughout this time, however, the work achieved within the organisation has been of high scientific merit.

Situated in central Yunnan, the area of Kunming is skirted on one side by the Dianchi Lake and on the remaining three by mountains. It therefore enjoys a sheltered microclimate, not unlike a temperate spring, avoiding the bitter cold of winter and the arid heat of summer otherwise typical of the area. The 100 acres that make up the Kunming Botanical Garden are only 12 kilometres away from downtown Kunming, offering a welcome green space for the lively city centre. Together with the nearby Heilongtan Park, a scenic resort famed for its ancient trees, it constitutes a significant area of natural reserve in the Yunnan province famed for its flora and biodiversity.

The Institute endeavours to engage the public simultaneously with its research and exhibitions, and the Garden is a perfect device for bridging this gap. Some 4,000 varieties of tropical and subtropical plants grow profusely amidst dense clusters of the official flower of Kunming, the camellia. The Camellia Garden is unique in the country and displays up to 40 species of flower. Ten special plant gardens and experimental zones arrange plants for closer inspection and the Tea Plantation in the Garden is one of the most distinct plantations in China. There are also specialised magnolia and begonia gardens, highlighting these popular ornamental plant varieties, as well as gardens of fern, hydrophytes and rare and endangered species. Conservation and genetic preservation is of great importance to the Institute, as is the study of plants used for industrial purposes, the export of Camphor oil and other plant volatiles being one of the major exports from the Yunnan Province. The Garden has reintroduced precious flowers, traditional medicinal herbs, important trees and endangered plants to the province, as well as developing technologies which help the migration and survival rate of plants transported from one region to another.

Kunming also participates in the largest flower auction, in all of Asia, the Kunming International Flower Auction at the world-famous Dounan Flower Market. The auction and market feature gigantic plant models fashioned from colourful blossoms, some of which were pioneered at the Institute's very own gardens. The organic combination of natural environment, research facilities and over 60 years' accumulation of academic resources have made it possible for the Institute to have developed its advanced research in Chinese botany, and the Kunming Botanical Garden allows an interaction with the public which makes it a comprehensive establishment that elegantly combines scientific research with education and pleasure.

tag: The Sweetgum Avenue, a part of the Hamamelidaceae Collection of Kunming Botanical Garden, which attracts thousands of visitors from all over China. opposite: View of the Cactus, Orchids, Fern, Subtropical and Propagation Houses. all photographs: © Kunming Botanical Garden.

SEAC PAI VAN PARK, MACAU

Situated in the wooded hillside of a small island off the Macau Peninsula, the Seac Pai Van Park is an idyllic location to take respite from the region's rapid urban development. Occupying 20 acres of forestry land on the comparatively undeveloped Ilha de Coloane, this park celebrates the natural landscape, flora and fauna of Macau, and encourages the Macanese people, as well as tourists, to familiarise themselves with the local ecology.

As the Macau peninsula underwent an exorbitant rate of development over the latter half of the twentieth century, governmental institutions began to realise the importance of rescuing some areas of cultural importance and natural beauty from the grips of industrialisation. In an effort to resist the encroachment of high-rise sprawl, Coloane and in particular the area occupied by Seac Pai Van was officially protected from development, and the park in its current form was inaugurated in 1985. A primary objective of the Institution is to sensitise the public to domestic and indigenous vegetation and wildlife, which are abundant and varied despite Macau's comparative lack of natural resources.

The Park features several facilities to encourage and facilitate public appreciation; broad and well-maintained paths lead visitors through landscaped areas and specialised gardens, and all botanical and zoological exhibits are clearly labelled in English, Chinese and Portuguese, reflecting the region's colonial legacy. As well as providing a popular area for the enjoyment of a natural environment, Seac Pai Van Park contains several specialised planting beds. The Garden of Medicinal Plants features 137 species that are used specifically in local indigenous and traditional Chinese herbal medicines, and function in conjunction with the Natural and Agrarian Museum to provide a historical perspective on the botanical culture of the region. The Museum seeks to preserve and educate visitors and the local population about traditional farming and ecology, maintaining techniques native to the area, and thereby going some way to preserving Macanese cultural history.

The Garden of Fragrant Flowers promises to be a sensuous treat, with much ornamental planting and landscaping is displayed emphasising natural floral perfumes. The Garden of Exotic Plants is a lush exhibition featuring a broad variety of colourful and exotic blooms and foliage native to the South China seaside and to the tropical and sub-tropical climates worldwide.

An Imported Plant Trial Zone has been implemented to research the success of foreign plants growing in the Macanese climate;

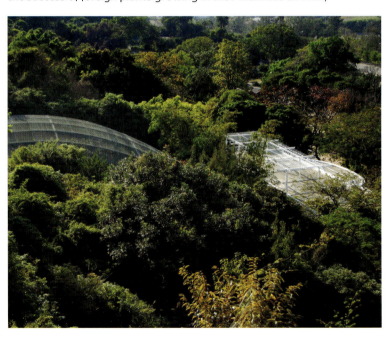

indeed, although the focus of the park is placed upon public education and enjoyment, a scientific research laboratory also operates from within Seac Pai Van, providing important research results to the international botanical community.

Seac Pai Van also contains a modest zoo, with a small spectrum of animal enclosures and exhibits. These vary from several species of wild monkeys, through to domesticated water buffalo, whilst a walk-in aviary offers the opportunity to view Macau's exotic bird

species, as well as more domestic breeds such as turkeys, geese and hens. These resources are especially important at school level education, as the children of the region have otherwise little exposure to agrarian or wild animals.

An arboretum leads off from the park and showcases over 100 species of tree; with a trail winding its way across nearly two kilometres, eventually arriving at a monumental and elegant sculpture of A-Ma, the Chinese sea goddess revered across several countries and cultures of the Far East and South East Asia. Several pagoda-like pavilions along the trail, as well as the summit, permit the visitor to enjoy panoramic views of the South China Sea coastline and local landscape.

tag: *Psychotria asiatica,* a common wild plant from Macau which can also be used for medicinal purposes. opposite: Aerial view of the Aviary. above: View of the Natural Education Walk. all photographs: Instituto para os Assuntos Cívicos e Municipais.

Spread over 173 acres, the Wuhan Botanic Garden is divided into several different landscapes with a variety of garden types throughout. The Aquatic Plant Garden is home to an impressive collection of conservation plants, set within a beautiful labrinthine display, culminating in a unique glimpse of rare underwater flora. Equally remarkable is the Rare and Endangered Plant Garden, which boasts the fruits of Wuhan's extensive conservation programme. The Ornamental Garden is found within a remarkable setting and offers a treat for the senses, complete with a stunning example of traditional Chinese architecture. Within its limits Wuhan also has a Chinese Gooseberry Garden, a Tree Garden, a Pine and Cyprus Garden and a Bamboo Garden.

Wuhan Botanic Garden has a rich history of conservation and education, emerging as the largest biodiversity protection base in all of China, and playing an important role in the conservation programme of the Wuhan and Hubei Provinces. Indeed, the Garden currently cultivates over 4,000 conservation species. Beyond this

instrumental role in the botanical health of its own region, Wuhan extends its influence internationally by organising important symposia, where over 40 countries have the opportunity to meet for a seed exchange and to discuss and debate botanical techniques. Wuhan specialises in plant propagation techniques and planning for afforesting activities, lending its expertise to botanic gardens and aboretums worldwide. Within its own limits Wuhan has selected and bred more than 40 new varieties of fruit trees, and has undertaken extensive forestation and cultivation of medicinal plants.

Beyond these comprehensive aims of conservation, Wuhan Botanic Garden also provides a beautiful respite from the otherwise stark Wuhan landscape, with serene lagoons and misty lakes dotted with lavish bursts of colour and exotic rock formations. Its aims of education are enhanced by the sheer aesthetic splendour of the grounds themselves, making this Botanic Garden a particularly special place in which to work and learn.

tag: Plant irrigation taking place in the Garden. opposite top left: The Rock Garden. opposite top right: The Fountain and Conservatory. opposite bottom: A bridge surrounded by tropical plants. below: The Garden's Aquatic Plant Collection.

COSTA RICA
ELSE KIENTZLER BOTANICAL GARDENS, SARCHÍ

Nestled at the foot of Volcano Póas in Costa Rica's central valley, Sarchi, Else Kientzler Botanical Gardens are spread across 17 acres, housing 2,000 plants from around the world. The Gardens form part of Innovaplant of Costa Rica, an agricultural organisation that exports ornamental plants and cuttings worldwide. Sarchí is known as a coffee town, and the Gardens began their life as a coffee plantation; the crop plants were removed, a nursery constructed, more than two kilometres of footpaths and trails were put in place eventually the Botanical Gardens were recognised as an area of cultural and scientific importance.

The reception building has been built in the style of a Costa Rican adobe villa, and this overlooks a labyrinth in the Hibiscus Garden, shaped like the flower after which it is named and inspired by a European tradition of geometrically designed mazes. Nearby, a water garden hosts floating waterlillies and Giant Marantas, alongside rare examples of papyri, the plant the ancient Egyptians used to make paper. An exotic fruit orchard features mangoes, mandarins and several unusual South American varieties, such as the Water Apple, amongst its collection of 40 types of fruit tree. Two of the most popular exhibits are the specialised orchid and hibiscus gardens, where dozens of examples of these two beautiful ornamentals can be admired year-round.

Helicons, one of the most recognisable tropical plants with their distinctive, spiky form, have their own designated garden at the Else Kientzler Botanical Gardens, and not only are they visually arresting, they also attract the attention of several species of hummingbird and the rare *Morphos peleides*, a species of large blue butterfly. Frogs are more frequent visitors to the Bromeliad collection, which features pinuella-type plants of varying colours and sizes. Here, in spring, they deposit their eggs in the water pouches formed by the plants, while in the dry season the trees in this section are filled with orange blossoms. The Gardens provide an ideal environment for succulents, plants suited to arid climates that cannot survive with limited sunlight, happily growing in the sunny climates of Costa Rica.

Near the middle of the Gardens is a unique rest area, with raised grass beds and benches for sleeping and reading—features that encourage the simple pleasures of enjoying the lush green environs of the Garden, whilst taking pleasure in the stunning views of the Poaz volcano and the spires of Church Sarchí below.

The site is bordered on the east by the River Trojas, and is surrounded by lush tropical vegetation that the organisation

insists must be protected at any cost. The Botanical Garden has a manifest emphasis on preservation, and cultivates many trees in danger of extinction; here you can find Red Foxwood (*Dalbergia retusa*), Zebrawood *(Astronium graveolens)* and Roughbark Lignumvitae (*Guaiacum sanctum*), as well as rare species of flowering groundcovers and shrubs.

tag: Passion Flower (*Passiflora*), a climber plant which attracts a variety of butterflies. photograph: Quijano. opposite top: The Heliconias Garden. photograph: Ludwig Kientzler. opposite bottom: A Pergola in the Bromeliads Garden. photograph: Ludwig Kientzler. above: Aquatic plants (*Pistia stratiotes, Nymphaea tetragona, Mayaca fluviatilis, Eichornia crassipes* and *Calthea lutea*) in the Else Kientzler Botanic Gardens Pond. photograph: Ludwig Kientzler.

CROATIA
ARBORETUM TRSTENO

The Arboretum Trsteno is located in the surroundings of Dubrovnik, in the small municipality of Trsteno, and covers an area of just over 61 acres. Trsteno, with its temperate climate, proximity to the sea and lush green surroundings was a natural destination for Dubrovnik nobility during the summer months. By the end of the fifteenth century, the site of the Arboretum was established as a park and summer residence of its patrician family, the Gučetić. In 1494 Ivan Marinov Gučetić began to design his house in a lavish Gothic style, and equipped his lands with several ornamental sculptures and stone features to complement his cherished collection of exotic cultivated plants, acquired from mariner friends travelling far overseas. He also established an aqueduct, to ensure a good supply of fresh water for the parkland, a supply that is still utilised today. The Arboretum Trsteno has a history spanning five centuries, and the aesthetic development can be mapped through Gothic, Renaissance Baroque, and Romantic forms; it is the oldest garden of its kind in this part of the world. The splendour of the Arboretum's architectural features and ornaments is rivalled only by its prestigious collection of trees, in particular, two splendid Oriental Plane trees. Rising to approximately 60 metres in height and five metres in diameter, the trees equal the majesty of the architectural features and follies in stature and in age, thought to be more than 500 years old, they are the pride of Trsteno.

The Arboretum is home to the fifteenth century summer house, its surrounding parkland dotted with sculptures of all styles. A chapel established by the first owner also stands here, alongside the later nineteenth century Neo-Romantic Drvarica Park, which hosts the larger part of the plant collection of over 300 species and cultivars. One of the dynastic additions to the Trsteno parklands is a Baroque fountain, featuring the mythological god Neptune surrounded by several nymphs. The last private owner, Vito Gučetić added romantic sculptures to the nineteenth century park, amongst them a statuette of Saint Nicole, as well as a formal topiary constuction cut from box hedge. The whole of the private estate was seized following the Second World War and has been the property of the Croatian Academy of Sciences and Arts since 1948.

In 1962, the whole area was declared a national rarity and registered in the list of protected natural monuments list as a protected site of landscape, preserving an area of approximately 255,000 square metres. Despite this, Trsteno suffered unprecedented and extensive damage and looting during the Yugoslav wars in the latter part of the twentieth century. In 1991, the Yugoslav People's Army launched a series of gunboat and air attacks and set the Arboretum alight, destroying large sections and causing partial damage to the summer residence as well as the oldest part of the Arboretum. The grounds were further damaged in 2000 by a drought-induced forest fire where 120,000 square metres succumbed to the blaze.

Hard work, and growing levels of tourism in the area, have helped the Arboretum to reclaim its former grandeur. Its wealth of plants and the extraordinary range of architectural styles featured in its summer manor, pavilion, aqueduct and Baroque fountain make this an unparalleled site of historic interest, and promise the ideal setting for the Garden's ongoing reconstruction.

tag and opposite top right: One of the Arboretum's many Cypress Trees (*Cupresus sempervirens 'Stricta'*). below: The Renaissance Garden. opposite top right: The Renaissance Fountain. opposite bottom: View of the Elaphite Isles from the Arboretum's Pavilion.

BOTANICAL GARDENS OF THE UNIVERSITY OF ZAGREB

In the centre of the city, a ten-minute walk from the Jelac'ic Square, sit the Botanical Gardens of the University of Zagreb. Its grounds were awarded to the University as a gift by the city authorities in the late nineteenth century on the condition that they would always be open to the public for the purposes of education and recreation. 115 years later, the University has not only kept its promise, but put years of devoted work into cultivating this ornate garden, exhibiting the full spectrum of Croatia's indigenous flora and fauna, which, regardless of the country's modest size, is remarkably diverse.

The Gardens were laid out in 1890, and first planted in 1892; experimental seeds were cultivated for research into plant physiology and virology. Greenhouses were built, as well as a director's building in the Art Nouveau style, an exhibition pavilion, and a Botanical Institute building, which was never completed. The plants grew astonishingly quickly and in just ten years all available greenhouse space had been used up, and this necessitated their closure to the public. Today they are still closed as the limited space allows only a few people to enter at any one time, but in spring many of the plants are transplanted outside for public view. During the Second World War, the Gardens were greatly damaged, including a marine aquarium dating from 1911. Throughout the hardships of the post-war years, some areas became dilapidated due to lack of funding, and some of the original glasshouses still show the scars of damage and disrepair. Careful and dedicated work ensured that the Gardens recovered their functionality and the University was able to continue its mission of housing all of Croatia's 5,500 species of ferns and seed-bearing plants.

The Gardens are, for the greatest part, designed in the style of a Croatian landscape, with freestanding clumps of trees constituting a dispersed arboretum, intersected by winding paths; this style of layout is synonymous with the University's stated desire to investigate the indigenous landscape of Croatia. Green lizards frequent the Gardens to bask in the sun, and bats, martens,

hedgehogs and dormice can also be spotted here. The Flower Parterre Section, which displays ornamental perennials and annuals, is designed in a formal French style, featuring flowerbeds with pleasing symmetrical lines.

Due to lack of space and funding, the goal of documenting the whole spectrum of Croatia's botanical biodiversity has not yet been reached. But the Gardens have not faltered in increasing the international collection and exhibits; by 1985, sub-Mediterranean, Mediterranean, Alpine and West European rock gardens had been established. A small laboratory is run for in vitro vegetative reproduction of plants and the germination of seeds, and plant collections are renewed by exchanging seeds with 300 gardens worldwide.

A stunning example of an acquisition through such an exchange is the *Victoria amazonica*, the largest waterlily species in the world. It was discovered in the Amazon in 1801, and successfully cultivated by Joseph Paxton 1849, in a greenhouse at the Duke of Devonshire's estate at Chatsworth. Its spiny leaves with raised edges can grow up to two metres in diameter, and its flowers, which are pollinated by nocturnal insects and open from dusk until dawn, can grow up to forty centimetres in diameter.

The Gardens continue to develop apace; recent additions and refurbishments include the restoration of the original nineteenth century wrought iron railings, and a domed glass structure built in 1996 brings the total number of greenhouses to 14. Future plans for the Gardens include the construction of new Mediterranean rock gardens and a small garden for medicinal herbs.

tag: Waterlily (*Victoria amazonica*). opposite top: View through woodland to the Gardens' lake. opposite bottom left: *Plumeria rubra*, with one of the Gardens' main buildings in the background. opposite bottom right: View overlooking the Gardens' Fountain. all photographs: Botanical Gardens, Faculty of Science, University of Zagreb, Croatia.

MENDEL UNIVERSITY OF AGRICULTURE AND FORESTRY
ARBORETUM AND BOTANICAL GARDENS, BRNO

The Mendel Arboretum and Botanical Garden grew out of an arboretum intended for forest engineers in the 1930s. In 1967 it became necessary, due to the expansion of the University, to create a new garden, and in 1970 nearly 25 acres was devoted to building a facility for the students of forestry and botany.

The Gardens are noted for its orchids, salix, cotoneaster and elaborate displays of alpine plants. The alpines—including a unique collection of Saxifraga—can be seen in a specially designed greenhouse complete with a controlled climate to replicate the habitat to which these plants are accustomed. In other areas of the Garden, perennials are grouped according to their place of origin, with plants from the Mediterranean, the Caucasus, North America, East Asia and the Southern hemisphere, occupying different beds along meandering walkways. In an area called the Botanical System, plants are grouped according to a fascinating display exhibiting their various medicinal or poisonous properties, including ancient examples of *Salvia officinalis*, *Mentha y piperita* and *Melissa officinalis*.

The plants in the Gardens are arranged according the Takhtadjan system. Takhtadjan was an influential soviet-Armenian botanist who developed a system of grouping flowering plants according to classes, orders and families, in a unique system that is employed variously worldwide, for example, in the Montreal Botanical Garden. Rare indigenous species can be found in the Native Flora Garden, which features endangered breeds such as *Gypsophila fastigiata* and *Dracocephalum austriacum*.

The Botanical Gardens also house a garden for the blind, with elevated flowerbeds and Braille signposts, and an emphasis on aromatic, prickly, tactile leaves and flowers, including water plants and woody species.

During the winter months, south-facing trench beds are lowered under the surface of the soil and covered, to insulate the flora and prevent them from succumbing to the frost. This ingenious solution was developed to protect these temperate plants from the hostile

Czech climate, allowing the Gardens to cultivate species from the Mediterranean, the Americas and Oceania.

The Garden of Miniatures, housing various saxifrages and dwarf species has been constructed, with plants growing out of a spectacular system of travertine arranged into a living sculpture that allows the visitor to have a close look at these subtle plants. The greenhouses contain orchids from the Americas and South East Asia, and the Rosarium displays several groups of winter-resistant botanical and park roses. One of the most impressive displays in the Gardens during summer months—when the ornamental grasses of the plant-beds are fully grown—is the collection of *Asteraceae*; an evolutionarily young family of plants that has become one of the most prominent and widely used in domestic gardening.

tag: The Iris Collection. above: The Garden for Blind People. opposite: The Ravine, a small deep valley comprising concrete blocks, which separate geographically arranged groups of plants. all photographs: Marketa Nohelova.

Founded in 1969 and opened to the public in 1992, Prague Botanical Garden is divided into several carefully landscaped open-air exhibitions. Also on the site is an impressive and unique greenhouse—Fata Morgana—that shelters much of the Garden's prized and extensive collection of internationally introduced flora. Rather than scientific advancement, these gardens are primarily dedicated to the sheer pleasure and admiration of plant life, and the result is a collection of expressive and stimulating environments.

The Japanese Garden is the larger of the open-air exhibits, and is itself divided into two sections; a lake forms the centerpiece of one side, with representational landscaping and planting. A turtle shaped islet and a careful stone "of Night Expectation" are a nod towards poetic Japanese symbolism, and the Garden strives to maintain a harmonious balance of focus between trees, grasses, water and mountain landscapes. Clumps of bamboo and plants indigenous to the Japanese landscape such as maples and irises continue to evoke the oriental influence, as does the Japanese Cherry Tree Grove, which forms the second part of this garden. Breathtaking in springtime blossom, there are over 26 species of cherry tree cultivated here, successfully conveying their Shinto-endowed qualities of subtlety and grace. A vigorous stream descending from the heights of the surrounding mountainscape courses around the area diverted through cascades and waterfalls, adding an aural element and a serene sense of continuity to the experience, this is accompanied by a small fountain to further add to the pleasing sounds and sights of the water feature.

The seasonal focus is continued into other parts of the Garden, in order to ensure that all times of year are enjoyable for the visitor; perennial blossoms are planted throughout the open-air gardens, as well as an impressive bulbous plant section with staggered flowering seasons. Near the lake, the autumn season of wild grasses and meadow saffrons, locally known as 'naked ladies', promise spectacular displays of colour.

The collection extends into a Mediterranean lawn terminating at a small alpine house, but another significant and undoubtedly popular area is the Garden's own vineyard. St Claire's Vineyard holds a national heritage status and is the largest of its kind in Prague; situated on top of a hillside and still in operation, it offers the opportunity to take in panoramic views over the church spires of the area and of the surrounding countryside, whilst sampling local and Moravian wines.

The Fata Morgana greenhouse is the prize of the Prague Botanical Garden, promising a 'tour of the world through the tropics'. Its unusual and innovative S-shaped design takes advantage of the rocky terrain surrounding the structure, and is divided into three truncated sections covering 1,750 square meters in total. Moving through the three differently climates houses within the greenhouse, one gradually becomes acquainted with the vegetation of tropical and of subtropical zones; plants are curated according to the phytogeographical regions and arranged to evoke natural formations. The first section displays succulents and other xeric or arid location plants, originating from Africa, Central and South America and Australasia. This flows into the humid lowland section, which is the largest exhibit and contains tropical and rainforest vegetation while nearby waterfalls and two aquariums replicate the lush habitats of Polynesia, Madagascar and many other tropical locations from which the plants originate. Unusually, this greenhouse terminates in a chilled climate, presenting hardy plants which thrive in the rough mountainous regions of the Andes, as well as mainland and island Asia and South Africa, with a focus on the rare vegetation of the Tepui mountains of Venezuela.

tag: The Mediterranean Section of the Prague Botanical Garden. opposite: The Tropical Section of the Prague Botanical Garden Greenhouse.

Joseph Hooker Stuart Gage Orto Botanico Padua Jardin des Plantes Chelsea Physic Garden Botanic Gardens Kew Kebun Raya Bogor Botanical Gardens Botanic Gardens Sydney Joseph de Jussieu Richardia grandiflora Botanical Garden Brooklyn Botanic Garden The Eden Project

The Changing Face
of the Botanic Garden

The Changing Face of the Botanic Garden

Brian Johnson

The Relationship between Plants and People

To walk into the herbarium at the Royal Botanic Gardens, Kew, in London, is to step into history. Stacks of dried, pressed and mounted plant specimens from around the world rise from worktables, as they are prepared for storage in tall metal cabinets that line the balconies and floors above. A distinct odour—a smell that can only be described as 'old'—pervades the room. Here, in this chamber, are plants from around the world, collected not just today or ten years ago—but also hundreds of years ago. More than seven million specimens fill these halls. The luckiest of visitors may be treated to a viewing of *Richardia grandiflora* collected in Brazil by Charles Darwin in 1832 or *Saussurea hieracioides* collected in India by Joseph Hooker in 1849. Some of these specimens were collected in the pursuit of scientific knowledge, others in pursuit of riches. The herbarium's cabinets are filled with countless stories of exploration, conquest, economic expansion, scientific advancement, and even greed.

These are also the stories of the changing face of botanic gardens throughout history. The Kew herbarium specimens from the 1850s were collected for vastly different purposes than the specimens collected in the past decade. What was at one time a rush to collect, identify and profit is now a rush to collect, identify and conserve. The stories of these specimens—the who, how, when, where and why they were collected—can tell us much about the missions and mandates of the world's botanic gardens.

Botanic gardens have always been about plants. But the history of botanic gardens offers a unique window into how we humans have used and valued plants in the past several centuries. In the sixteenth century, plants were highly prized for their medicinal properties. 100 years later, the expansion of economic empires was driving the quest to find new plant species, and thus new sources of wealth, from around the world. By the end of the nineteenth century, the Industrial Revolution had stimulated great interest in the aesthetic values of landscapes and the plants that give them texture and colour. And today, scientists race to preserve tens of thousands of plant species before they become extinct. These pursuits have framed the evolution of botanic gardens. Whether plants were viewed at a moment in history as sources of medicine, income, pleasure or ecological stability can help us understand the role of botanic gardens at that same time.

In a 1937 article, C Stuart Gager, a longtime director of Brooklyn Botanic Garden, pointed to five overarching purposes for botanic gardens: science, education, recreation, civics, and economics.[1] Since the founding of the first botanic garden

A typical herbarium sheet, this spectacular specimen of *Gentiana sino-ornata* was collected by George Forrest in 1910, Image courtesy of the Royal Botanical Gardens, Edinburgh.

An example of early Egyptian gardens as represented in *Neb Amun's Gratitude for his Wealth*, a tomb painting dating back to 1415 BCE.

in Italy during the sixteenth century, every botanic garden has embodied each of these purposes to greater and lesser degrees. Whether a particular garden was driven more by science, education or economics was largely a product of an era's politics and cultural attitudes toward the natural world and plants, in particular. "Botanic gardens are phenomena of the cultures of nations and, as such, reflect dominant national interests in their patterns of growth", wrote Joseph Ewan in a 1972 essay for the Morton Arboretum.[2] By looking at botanic gardens in this context, we can begin to understand how they have evolved from small gardens growing medicinal plants during the sixteenth century to international institutions at the forefront of biodiversity conservation and education today.

An early engraving of the Botanic Garden at the University of Padua.

Physic Gardens: the First Botanic Gardens

Gardening, of course, is nothing new. The ancient Egyptians planted gardens to honour royalty, and the Aztecs were known to have planted medicinal gardens.[3] In China, the Shennong Garden of Medicinal Plants grew plants for scientific purposes more than 2,800 years ago.[4] But botanic gardens are a relatively recent phenomena in human history. A botanic garden is defined as an institution holding documented collections of living plants for the purposes of scientific research, conservation, display and education. It is the combination of these four purposes with a 'documented collection' that makes the botanic garden a recent development in history. Research is beginning to point to gardens that may have fit this definition thousands of years ago. Current research and literature, however, establishes the age of botanic gardens at just over 450 years.

The Orto Botanico Padua is widely recognised as the oldest botanic garden in the world. Begun in 1545 by the medical faculty of the University of Padua in Italy, the Garden still thrives today, albeit slightly expanded and with an evolved mission. What set this garden apart from its very beginning—what made it a *botanic garden*—is that the garden was a definite enclosure with plants grown and categorised for scientific, research and education purposes.

The Orto Botanico Padua was what is known as a physic garden. The word 'physic' refers to the 'healing arts' or what we would now call medicine. Physic gardens were cultivated for the study and growing of medicinal plants in order to reduce errors and fraud in creating pharmaceutical 'simples', or medicines.[5] Dishonest, fraudulent or poorly trained apothecaries were known to prescribe incorrect simples, resulting in a con in the best of cases and death in the worst.[6] The early physic gardens were most often affiliated with universities, and the gardens themselves were overseen

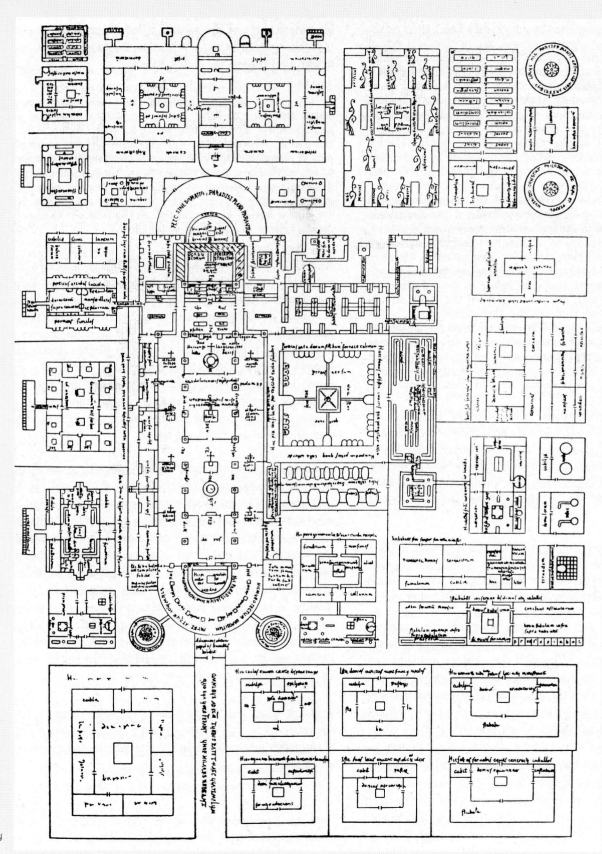

Some of the world's oldest proto-botanical gardens were housed within the walls of religious institutions like this plan of a Benedictine Monastery in St Gall, Switzerland.

by professors of medicine. It was not uncommon for a professor to teach a course in human anatomy during one term, and then to teach a course in botany the next.[7] The rise of physic gardens throughout Europe demonstrates the importance of plants as medicine during the sixteenth century. Dozens of gardens were established during this time, including physic gardens in Bologna, Italy, 1567; Leyden, The Netherlands, 1587; Paris, 1597; and Oxford, England, 1621. These gardens also played important roles as teaching gardens for university faculty and students, and their arrangements usually reflected this purpose. Plants at the physic garden in Pisa, Italy, for example, were grouped "according to their properties and morphological characteristics: thus one finds beds for poisonous plants, prickly plants, smelling plants, bulbs and marsh plants".[8] An efficient system of organisation for a garden made it a more useful educational tool.

It wasn't long, however, before these physic gardens expanded their collections to include not just medicinal plants, but any kind of plants. A writer in 1561 noted that "in botanic gardens not only medicinal herbs were cultivated but also other plants, especially rare ones, for the purpose of observing and admiring nature".[9] Plants in botanic gardens were coming from other countries in Europe, as well as Egypt, Syria, and other distant places. Collections in gardens were booming. At Paris' Jardin des Plantes, for example, 1,800 species were recorded in 1636, 2,360 species in 1640, and 4,000 species in 1665. An intense rivalry sprung up between gardens to obtain the largest collections from around the world.

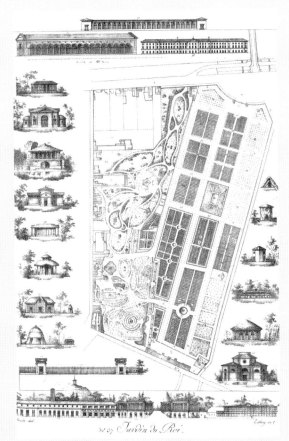

An 1820 plan of the Jardin des Plantes in Paris, the former Jardin du Roi is the oldest botanic garden in the French capital.

Activities at other botanic gardens during the seventeenth century also point to a shift in the role of botanic gardens, away from an exclusive focus on medicinal plants. Thomas Baskerville, in his *Account of Oxford Collectanea*, 1670–1700, highlights some of these new purposes at the University of Oxford Botanic Garden in England:

> Amongst ye severall famous structures & curiosities wherewith ye flourishing University of Oxford is enriched, that of ye Publick Physick Garden deserves not ye last place, being a matter of great use & ornament, prouving serviceable not only to all Physitians, Apothecaryes, and those who are more immediately concerned in the practice of Physick, but to persons of all qualities seruing to help ye diseased and for ye delight & pleasure of those of perfect health, containing therein 3,000 seuerall sorts of plants for ye honor of our nation and Universitie and service of ye Commonwealth.

Here we find some of the first mentions of botanic garden activities and roles that would soon outstrip medicine of its primary importance: gardens as places of national prestige, and gardens as places of personal pleasure.

Other gardens were evolving in their role and functions as well. The Chelsea Physic Garden, established in 1673 by the London Society of Apothecaries, was charged in 1722 to deliver 50 new dried and preserved plant species every year to the Royal Society of London for 40 years. This mandate by the Garden's main benefactor ensured "that 2,000 different species of trees, shrubs and flowers would be grown in the Garden during that time".[10] The benefactor believed that the Physic Garden's purpose should include not just the growing of medicinal plants, but also the discovery and identification of new plant species from around the world. His order was thus a condition of his patronage.

In all of these seemingly small changes and expansions, the face of botanic gardens was undergoing a major transformation. While people still relied on plants for medicine, new value was beginning to be placed on plants as sources of financial opportunity and aesthetic enjoyment. As human interest in plants evolved, so did the role of botanic gardens.

Kew Green with the road leading to Kew Palace, mid-1760s drawing by George Bickham.

Botanic Gardens in the Age of Empire

By the eighteenth century, the rulers of the great empires of Europe—in England, France, The Netherlands, and Spain—were in a global race to find new riches. These kings and queens sponsored expeditions to all reaches of the earth in the hope of finding new lands and peoples to rule, and new products to export. What were these new opportunities for wealth? Often, they were textiles, china, gold and silver. But of equal importance were plant species from around the world. It was during this age of empire that botanic gardens in the Americas, Asia and Africa had their beginnings. More than 100 gardens including the Indian Botanic Garden (India) and Peradeniya Botanic Garden (Sri Lanka) can trace their roots to their roles as outposts for the cultivation and eventual transport of economically important plants. A few of these gardens would eventually rise to become some of the greatest botanic gardens of the world. For the time being, however, economic expansion in the botanical arena was driven by gardens in the old world.

Sir Joseph Banks, respected advisor and contributor to the Royal Botanic Gardens at Kew.

As the physic gardens of Europe began to expand their collections and therefore their missions, kings, queens and other powerful individuals began to take note of these botanic gardens. In 1759, for example, England's Princess Augusta, Princess Dowager of Wales, acted upon her longtime interest in botany and ordered the creation of a physic garden as part of the Royal Garden attached to Kew House outside of London. In subsequent years, Augusta built glasshouses and an arboretum. Upon Princess Augusta's death in 1772, King George III combined the gardens at Kew with those of the Palace of Richmond, forming what is now known as the Royal Botanic Gardens, Kew. The important role this new botanic garden would play in the British Empire's future was underscored by Sir Joseph Banks' appointment as botanical adviser to the King and as first unofficial director of the Royal Botanic Gardens, Kew. "The days of Sir Joseph Banks were indeed the Golden Age of Kew", writes Arthur Hill in *The History and Function of Botanic Gardens* "and under his direction the Royal Gardens became a centre of botanical exploration and horticultural experiment unparalleled before or since".

It was the choice of Banks as botanical advisor and garden director that turned Kew, and perhaps the world of botanic gardens as a whole, into the botanical centres they would later become. Banks was an experienced explorer, a dedicated naturalist, and president of the Royal Society, London. He had sailed and explored the Pacific Islands, as well as New Zealand and Australia, with Captain James Cook. As his involvement with Kew grew, he became an advocate for expeditions around the world in a quest to make Kew the world's pre-eminent botanic garden. "In order to establish Kew's superiority to gardens at home and abroad, Banks insisted that "as many of the new plants as possible should make their first appearance at the Royal Gardens". It became a matter of personal pride to outstrip rival botanical gardens in Vienna and Paris.[11]

During his decades of work at Kew, Banks ordered plant identification labels to replace the numbering system originally in use at Kew. He encouraged any associates who were embarking on a foreign voyage, whether for business, government, military or other purpose, to bring back seeds or plant specimens for

Kew. These merchants, missionaries and foreign emissaries brought new plants to Europe for the first time, including a Chinese Tree Peony and *Hydrangea hortensia*. "The vision of botanical gardens participating in the commercial life of the nation came to fruition with Banks' transformation of Kew into a centre for the global transfer of plants through its collectors and links with colonial botanical gardens."[12] Kew quickly became the centre of economic botany for the British Empire.

The plants that created this economic power are common around the world today. During the eighteenth and nineteenth centuries, however, new plants were regularly 'discovered' and botanic gardens played a subsequent role in exporting these new cash crops around the world. A few examples demonstrate this point.

Today, billions of people around the world start their day or finish a meal with a cup of tea. The plant's origins, however, are in China, which dominated the tea industry into the nineteenth century. The British were eager to establish their own stake in the cultivation of tea, and in the first half of the nineteenth century made several attempts to cultivate tea in their colonial botanic gardens in India. These efforts met with little success. It wasn't until 1851 that a British trader was able to export 2,000 tea plants and 17,000 tea seeds, and recruit several tea growing experts from China. This was enough to begin mass cultivation in the Indian regions of Darjeeling and Assam. Soon, Indian tea had supplanted Chinese tea on British tables and this transformation would not have been possible without the work of the British colonial botanic gardens in India.[13] Tea also went on to be introduced to Ceylon (now Sri Lanka) via the Peradeniya Botanic Garden. In the ten years between 1876 and 1886, tea planting in Ceylon expanded from 2,000 acres to 100,000 acres, and the Botanic Garden was "unquestionably responsible for Ceylon's very profitable tea industry".[14]

Early example of a Chinese tea plantation. China dominated the tea industry well into the nineteenth century.

Europe had known of the wonders of rubber for hundreds of years before its wide distribution by botanic gardens during the nineteenth century. Indeed, "the Spaniards reported seeing bouncing balls in Mexico", and the "French explorer La Condamine brought the first specimens of *caoutchouc* to Europe in the eighteenth century".[15] The rubber industry, however, relied on wild crops from the Amazon well into the nineteenth century. In 1876, a plant collector commissioned by Kew received permission from the Brazilian authorities to export 70,000 seeds to England. One year later, nearly 2,000 plants were en route to the Peradeniya Botanic Garden in Ceylon and from there to botanic gardens throughout the British Empire in Asia. The Singapore Botanic Gardens received 22 of these plants, and over the next several decades distributed seven million seeds throughout the region, leading to the successful establishment of rubber plantations throughout the colonies.

Malaria is viewed today as a tropical disease. In the seventeenth and eighteenth centuries, however, malaria was common in European states as well. Even King Charles II of England was afflicted with the disease. The presence of the disease on the European continent, along with European economic interests and colonies in tropical areas of Asia, Africa and the Americas, led to worldwide demand for malaria's first known cure—the bark of the Cinchona tree, native to the Andes

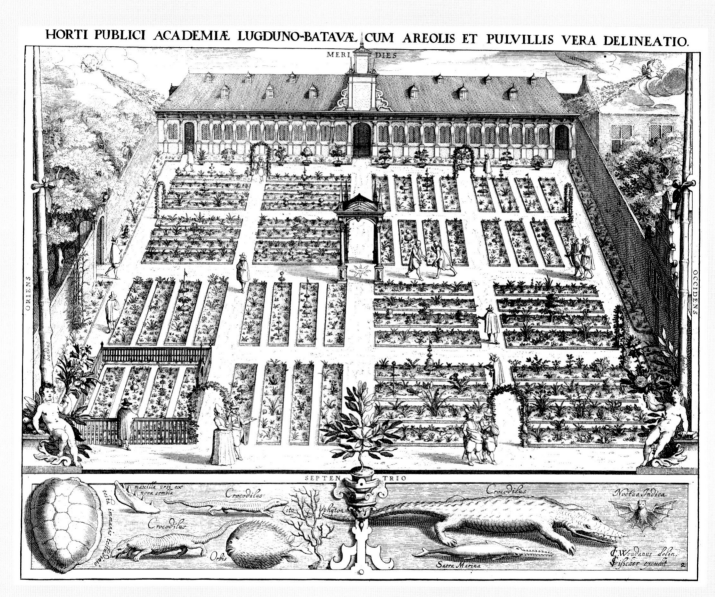

Historic engraving of the Hortus Botanicus Leiden, the oldest botanic garden in The Netherlands and frequent host to famed botanist Carolus Linnaeus.

Mountains of South America. Early efforts to smuggle the plants or seeds from South America for cultivation in European colonies in Asia failed when the plants were unable to survive in their new surroundings. By 1860, South American exports of Cinchona bark had reached two million pounds annually, rising to nearly 20 million pounds in 1881. By 1884, however, that level had fallen to some four million pounds due to competition from cultivated crops in Asia. By this time, the British and Dutch were both successful in establishing Cinchona plantations in their Asian colonies. The British botanic gardens in India and Ceylon facilitated the planting of millions of trees in these colonies. The Dutch, however, through their botanic garden at Buitzenborg (now the Kebun Raya Bogor) on Java (Indonesia), eventually became the undisputed leader in Cinchona exports, capturing 80 per cent of the market.

The distribution throughout the European empires of these and other important crops, including coffee, pineapple, cocoa and spices, was one of the primary missions of the botanic gardens established outside the European continent. The presence of many botanic gardens with rich scientific traditions in parts of Asia, Africa, and the Caribbean can be traced to this era of economic expansion. At the turn of the twentieth century, 102 botanic gardens (or 38 per cent of the world's botanic gardens) were colonial gardens of the British Empire. Other European states controlled 18 colonial gardens during the same time. Colonial gardens such as the Kebun Raya Bogor (Indonesia), the South African National Botanical Garden, the Royal Botanic Gardens Sydney (Australia), and the Singapore Botanical Gardens are today considered some of the great botanic gardens of the world.

Gardens during the Golden Age of Botany

While plants helped fuel economic expansion during the eighteenth and nineteenth centuries, this era would also come to be regarded as the golden age of botany, and botanic gardens were at the centre of this magnificent time in the history of science. At the turn of the eighteenth century, physic gardens were amassing large collections of plants from around the world, and botanists were using competing systems to name and organise these countless new species. Names of plants were horrifically long, meaning that no person of average memory could ever learn the names of more than a hundred or so plants. Furthermore, there was no clear consensus on how to organise and categorise the myriad plant species of the world. The field of botany was, in essence, a chaotic mess.

This was the botanical world that Carolus Linnaeus (born Carl Linne), the man who would eventually transform the field, entered into as a student in the early eighteenth century. Linnaeus was a Swede who, during his studies of medicine at the University of Uppsala, became fascinated by the world of plants. He participated in plant collection trips in Sweden and eventually in other parts of Europe. After finishing his degree in The Netherlands, he spent time at botanic gardens there, notably the University of Leiden Botanic Garden. It was here that Linnaeus perfected and published the theories of plant classification that he had been developing for many years. His classification system was "widely acclaimed

Early nineteenth century view of the Singapore Botanic Garden, which started out as a British colonial garden and evolved into one of the most important independent gardens in the world.

Indonesia's Kebun Raya Bogor began its life as a Dutch colonial garden.

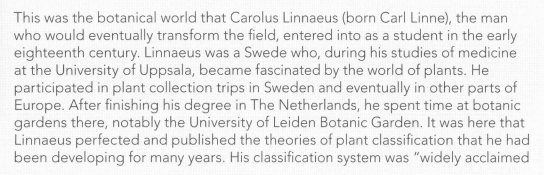

because his approach was so unusual and commendably simple; his system asking no more from a botanist than that he count the numbers of a flower's stamens. The total, noting certain other characteristics, would place the plant first in a division and then in a subdivision. Identification was as easy as that."[16] The classification system was an imperfect and artificial one, but it became the foundation for more technical and precise systems that evolved in later years. Linnaeus, however, is perhaps best known for the simplified naming system he created. Every species was to be given just a two-word Latin name, a system that has lasted to the present day.

Of course, Linnaeus' ideas and new systems were not universally accepted from the start. Some botanists objected to the sexual nature of his classification system, abhorred by references to plant stamens as 'bridegrooms', styles as 'brides', and plant reproduction as "celebrat(ing) their nuptials".[17] "This caused one plant to have '20 husbands or more in the same marriage'"—an idea and language that were far too risqué for the times.

While Linnaeus was revolutionising the science of botany, it was the directors and associates of botanic gardens that were actively expanding the world scientific community's known flora. In the early years of botanic gardens, plants and seeds were brought back as 'bonuses' from expeditions whose primary purposes were geographic exploration or religious missions. France's Louis XIV, for example, paid commissions to missionaries who brought back seeds and plants for the Jardin du Roi. Abbe Armand David, in the nineteenth century, "identified 52 new varieties of rhododendron, and more than 40 primulas" for the botanic garden.[18]

The famed botanist, Carolus Linnaeus. dressed in traditional Northern Swedish attire.

As economic botany rose in importance, however, so did the place of botany in global expeditions. At Kew in London, Joseph Banks dispatched a cadre of trained plant collectors around the world to bring back new and important species for the botanic garden. He "continually implored plant collectors to be on the lookout for potentially useful plants".[19] One of his collectors, Francis Masson, collected plants for more than 30 years to be brought back to Kew from Africa, the Caribbean, and North America. Amongst the many plants he collected were 50 varieties of Cape Pelargoniums, which serves as the ancestor to the common garden geranium so loved by the Victorians in England.

Botanic gardens in other nations similarly organised their own collecting expeditions. Joseph de Jussieu, in the service of France's Jardin des Plantes, was the first to collect Cinchona seeds and plants from the Peruvian Andes.[20] This 'discovery' along with the important preparation information he learned from the indigenous people there, would eventually lead to a worldwide push to fight malaria.

By the end of botany's golden age, the botanical sciences were well-ordered and well-respected within the scientific community. Meanwhile, botanic gardens, through their advances in economic botany and in plant science, were firmly rooted as the premier institutions of the field. But these botanic discoveries were also resonating with the average citizen—those outside the realms of science and economics. The general public was becoming enamored with the new flora of the

Various glasshouses in Belgium belonging to the early botanical artist Louis van Houtte.

world, and the joy and delight that flowers and plants could bring to their lives. Botanic gardens, whose glory had traditionally come from scientific discovery, were now being discovered by the public for the visual treasures they held inside their gates.

Gardens of Leisure

During the nineteenth century, cities in Europe and the United States were rapidly expanding under the economic explosion of the Industrial Revolution. Rural dwellers and new immigrants were drawn to urban areas to seek jobs and, they hoped, a better standard of living. Throughout the course of the nineteenth century, this unprecedented urban growth continued. In England and Wales, the percentage of the population living in cities greater than 10,000 skyrocketed from 21 per cent in 1801 to 62 per cent by 1890. Similarly, while just ten per cent of the population of England and Wales lived in cities greater than 100,000 at the turn of the nineteenth century, that number had risen to 33 per cent in 1891. Across the Atlantic Ocean in America, the situation was no different. The percentage of Americans living in cities greater than 10,000 rose from three per cent to 28 per cent, and in cities greater than 100,000 from zero per cent to 15 per cent.[21]

The cities that these new urban residents called home were crowded and oftentimes overwhelming places, and it was these conditions that gave rise to the urban parks movement in the nineteenth century. "Public parks, providing green refreshment for the bodies and souls of city-dwellers, were becoming a necessity for every well-run city."[22] In New York City, for example, city planners foresaw continued urban growth and set aside a large tract of land as a green space for public leisure and recreation. Central Park, as this area would come to be called, would become one of the most important parks in the United States, and its designers, Frederick Law Olmsted and Calvert Vaux, would become the most important landscape architects of the era.

This rapid expansion of cities also coincided with the expansion of plant collections at botanic gardens. As new species were brought from around the world to botanic gardens, public interest grew in viewing and in some cases cultivating these plants. By the year 1900, one million visitors were being attracted each year to the floral sights of the Royal Botanic Gardens, Kew. Throughout Europe, new botanic gardens were being established, "and nearly all of these gardens were mainly pleasure gardens with very few of them having any scientific programmes". Improved access to public transportation in cities allowed more and more urban dwellers to make an excursion to a botanic garden where they were transported to a green wonderland full of rich colours and sweet scents. As they strolled the tree-lined paths and meticulously laid out gardens, visitors were able to leave the crowds and noise of urban life behind.

Botanic garden visitors during the nineteenth century were also drawn to the new gardens under glass. Inside these glasshouse conservatories, made possible by advances in glass and cast iron technology, plants from tropical and arid environments flourished, the likes of which few had ever seen. These plants became spectacles that attracted visitors from the growing middle class, but greenhouses were nothing new to the botanical world. The origins of these 'hothouses' can be traced back to the sixteenth century, when they were developed to house less hardy plants during winter months in northern Europe. The glasshouses of the nineteenth century, however, served more than just a practical purpose. These towering architectural sculptures of iron and glass were sights to behold in and of themselves. "The interest and delight in the exotic could now be celebrated through the architecture of the enclosure, impressive enough to dignify the magnificence of the plant collection."[23] The glasshouses of the 1800s could also "be seen as fulfilling a desire for paradise that had risen out of fears of the rapid growth of industrialisation and the appalling conditions that came with it".[24]

The glasshouses of botanic gardens continue to draw visitors even today. At Paris' Jardin des Plantes, a glasshouse conservatory from 1836 is likely the oldest still in use. Hundreds of visitors to the Royal Botanic Gardens, Kew, snap photographs every day of the famed Palm House, constructed between 1845 and 1848. Even modern glasshouses, such as the 1960s geodesic dome Climatron at the Missouri Botanical Garden, continue to delight visitors with displays of exotic plants from around the world. Botanic gardens worldwide still aim to be a source of inspiration, relaxation and wonder for visitors of all ages.

The common Garden Geranium, whose ancestry can be traced back to over 50 varieties of Cape Pelargoniums acquired by Francis Masson on one of his many collecting expeditions.

Botanic Gardens Today: the Conservation Garden

The Beijing Botanical Garden is, like the city itself, a bustling place. Swarms of schoolchildren run along the meandering paths of the Garden, which is tucked neatly into the foothills on the ever-expanding edge of the city. Electric trams shuttle groups of visitors from the Garden's famed crabapple tree collection to its futuristic-looking conservatory. Scientists spend long hours in laboratories, researching plant genetics and consulting with biologists to plan species reintroduction efforts. This is the modern botanic garden, and China is at the forefront of new botanical endeavors and research. More than 100 new botanic gardens have been built in China since 1950, with dozens more in the building and planning stages. These gardens represent the current face of botanic gardens: conservation.

In the past several decades, the world has turned its attention to one of the defining issues of our modern era: the global environment. Our planet is facing unprecedented challenges, from biodiversity loss to deforestation to climate change, all linked to our activities as human beings. World leaders are debating solutions to many of these issues, and nongovernmental organisations worldwide have stepped up to provide resources and expertise in environmental decision-making.

An example of the *echinocactus*, the beautiful Barrel-Cactus, a genus whose species have dwindled down to less than six is just one of the many varieties of flora protected in modern conservation gardens.

Plants have not escaped these human-induced pressures. Botanists estimate that 100,000 plant species around the world—one-third of the total known global flora—are threatened with extinction, making plants the most endangered species on the planet. Habitat loss, invasive species, wild collection and climate change are all contributing to this global decline in plant diversity and plant populations.

And so now, centuries after the first botanic gardens were established to cultivate plants to fulfill a human need for safe medicine, the work of botanic gardens rests on an even simpler premise: humans and all species need plants in order to survive. The conservation work that is now driving the mission of the more than 2,000 botanic gardens around the world is summed up in the public awareness campaign of Botanic Gardens Conservation International, a worldwide network of botanic gardens: "Plants for Life".

The conservation agendas of botanic gardens around the world differ depending on the local environment, funding, government support, and cultural considerations, amongst other factors. Most botanic gardens, however, do engage to some degree in each of the following three conservation arenas: environmental education; science and research; and ex situ conservation projects.

Education has long been a key component of botanic gardens. The early physic gardens were in essence 'teaching gardens'—resources for university professors and students alike. Today, at most botanic gardens around the world, schoolchildren, adults, families and even professional horticulturists take part in education programmes almost daily. At the Limbe Botanical Garden in Cameroon, for example, children from local schools visit the Garden for hands-on activities to learn about their local environment. They plant trees and learn how to grow plants in a special garden designated for children. The Garden also provides visitors with guided tours, as well as a library and information centre.

At Brooklyn Botanic Garden in the United States, the Garden helps run a local secondary school with a focus on science and the environment. Students at the school perform many of their biology labs at the Garden with staff educators and scientists. And at botanic gardens such as England's Eden Project, environmental education is at the core of the institution's mission. The Garden was literally created as a place for education, and every aspect of the visitor experience is treated as a learning opportunity, from the on-site food service featuring local produce and waste-free meals, to the gift shop selling sustainably produced plants for home gardens.

All of these programmes, and similar programmes at botanic gardens around the world, promote the importance of plants and the need for their conservation. With more than 150 million visitors each year, botanic gardens are an ideal venue for communicating this message.

China's 30,000 known species of plants make up approximately ten per cent of the world's flora.[25] As the country moves toward rapid industrialisation and urbanisation, the Chinese botanic gardens are working to document and cultivate these plants. By 1994, 50 per cent of China's known flora was being cultivated in Chinese Academy of Science Botanic Gardens. Additional scientific mandates for Chinese botanic gardens include studying growth patterns and genetic variations of plant species, researching adaptability of introduced plant species, and assessing economic value of species.

China is not alone. Botanic gardens around the world have active science and research departments. At Cibodas Botanic Garden, Indonesia, for example, researchers cultivate and research dozens of types of moss in a special moss garden, which thrives in the Garden's cool, moist, mountainous environment. And the New York Botanical Garden's International Plant Science Center researches the evolutionary relationships of plants.

Additionally, the discovery of new plant species is not over, and botanic garden researchers are continuing to contribute to our understanding of the world's flora. The research function of botanic gardens is also extremely important in many countries, including the United States and the United Kingdom, where the botanical

Contemporary view of the Kirstenboch National Botanical Garden in South Africa.

The spectacular view onto the roof of the glasshouse from the Palm House at the Royal Botanical Gardens, Edinburgh.

sciences are losing favour at universities more interested in biotechnology and other 'economic' sciences of the modern era.

With an estimated one-third of plants around the world threatened with extinction, botanic gardens are racing to protect these imperiled species. In 2000, the *International Agenda for Botanic Gardens in Conservation* was developed to help guide the conservation work at botanic gardens around the world. The *Agenda* outlines more than 150 actions botanic gardens can take to integrate conservation as a priority at every level of the institution. One of the strategies highlighted by the *Agenda* and by global biodiversity conservation treaties such as the Convention on Biological Diversity is the implementation of ex situ conservation projects.

Ex situ conservation is designed "to provide protective custody. It is justifiable only as part of an overall conservation strategy to ensure that species ultimately survive in the wild."[26] It differs from in situ conservation, which aims to protect "biodiversity within ecosystems or natural habitats". Botanic gardens are the principal centres for the ex situ conservation of plant species. The world's botanic gardens hold more than four million plant specimens, representing nearly 100,000 plant species. These plants, along with seeds being held in seed banks at botanic gardens and research centres, may be important sources for future species reintroduction efforts in the wild. The *Global Strategy for Plant Conservation* mandates that 60 per cent of the world's threatened plant species be held in ex situ collections by the year 2010, and botanic gardens are leading this charge.

The Future Face of Botanic Gardens

The current plant conservation crisis is so extreme that it is difficult to predict the future for botanic gardens. As the full extent of plant species loss becomes known in the next several decades, botanic gardens will be key players in planning habitat restoration and species reintroduction efforts. In many parts of the world, where urbanisation is happening at a rampant pace, botanic gardens will continue to offer city residents respite from traffic, crowds, and pollution. And as our understanding of ecology continues to evolve, botanic gardens will continue to be important sources of information about the world's plants—the backbone of the earth's ecosystems. "In the context of contemporary concerns about climate change and damage to ecosystems… environmentally utopian visions remain very potent and the botanic garden contributes toward visualising them."[27]

No matter what the future face of botanic gardens may look like, one thing is certain: botanic gardens will continue to be humanity's main scientific, aesthetic and social link to plants. The foundations of the first botanic gardens nearly five centuries ago are little different than they are today. They serve to educate, to explore, to fascinate, and to discover. They will continue to reflect our evolving relationship with plants and the rest of the natural world, and they will continue to remind us of the many wonders of life here on earth.

1   Gager, CS, "Botanic Gardens of the World: Materials for a History", *Brooklyn Botanic Garden Record*, XXVI(3), 1937.

2   MacPhail, I, *Hortus Botanicus: The Botanic Garden & the Book: Fifty Books from the Sterling Morton Library exhibited at the Newberry Library for the Fiftieth Anniversary of the Morton Aroboretum*, Chicago: The Morton Arboretum, 1972.

3   Hyams, E and W MacQuitty, *Great Botanical Gardens of the World*, New York: The Macmillan Company, 1969.

4   Chengyih, W and T Fenqin, eds., *The Blossoming Botanical Gardens of the Chinese Academy of Sciences*, Beijing: Science Press, 1997.

5   Reeds, KM, *Botany in Medieval and Renaissance Universities*, New York: Garland Publishing, 1991.

6   Reeds, *Botany in Medieval and Renaissance Universities*.

7   Reeds, *Botany in Medieval and Renaissance Universities*.

8   Hill, AW, "The History and Function of Botanic Gardens", *Annals of the Missouri Botanical Garden*, 2 February, 1915.

9   Hill, "The History and Function of Botanic Gardens".

10 Drewitt, FD, *The Romance of the Apothecaries' Garden at Chelsea*, third edition, Cambridge: Cambridge University Press, 1928.

11 Desmond, R, *Kew: The History of the Royal Botanic Gardens*, The Harvill Press, 1998.

12 Desmond, *Kew: The History of the Royal Botanic Gardens*.

13 Brockway, LH, *Science and Colonial Expansion, The role of the British Royal Botanic Gardens*, New York: Academic Press, 1979.

14 Hyams and MacQuitty, *Great Botanical Gardens of the World*.

15 Soderstrom, M, *Recreating Eden: A Natural History of Botanical Gardens*, Vehicule Press, 2001.

16 Whittle, T, *The Plant Hunters*, New York: PAJ Publications, 1988.

17 Soderstrom, *Recreating Eden: A Natural History of Botanical Gardens*.

18 Soderstrom, *Recreating Eden: A Natural History of Botanical Gardens*.

19 Stern, W L, "The Uses of Botany, with Special Reference to the Eighteenth Century" *Taxon* 42(4).

20 Soderstrom, *Recreating Eden: A Natural History of Botanical Gardens*.

21 Blumin, SM, "Driven to the City: Urbanisation and Instustrialisation in the Nineteenth Century", *OAH Magazine of History*, 20(3), 2006.

22 Soderstrom, *Recreating Eden: A Natural History of Botanical Gardens*.

23 Baker, K, "Tempering the Elements: Botanic Gardens and the Search for Paradise", PLEA 2006, The 23rd Conference on Passive and Low Energy Architecture, Geneva, 2006.

24 Baker, "Tempering the Elements: Botanic Gardens and the Search for Paradise".

25 Chengyih and Fenqin, *The Blossoming Botanical Gardens of the Chinese Academy of Sciences*.

26 Wyse Jackson, PS and Sutherland, LA, *International Agenda for Botanic Gardens in Conservation*, Botanic Gardens Conservation International, 2000.

27 Baker, "Tempering the Elements: Botanic Gardens and the Search for Paradise".

DENMARK
BOTANIC GARDEN OF COPENHAGEN UNIVERSITY

This Botanic Garden is an Institute within the University of Copenhagen that cultivates a large collection of living plants for research and education, increasing botanical knowledge and awareness of nature, not only nationally but also on a global scale. The Garden displays Denmark's largest collection of living plants and houses the only gene bank for wild species in the country.

The first botanic garden of the University of Copenhagen, Hortus Medicus, was founded in the central part of the old city in 1600 AD. It was established on 2 August 1600, by royal charter and donation of land near Skidenstraede (now Krystalgade), which previously belonged to the Zoological Museum. A residence with an adjacent botanic garden plot was also built for one of the professors. No financial means for the maintenance of the Garden were provided for in the charter, and it was not until about 100 years later, in 1696, that one of the Garden's supervisors, Rasmus Caspar Bartholin, set up an endowment to pay for a permanent gardener. One of the most prominent men of his time, Ole Worm, 1588–1654, attempted to reform the teaching of medicine and botany around 1620 and revived plans to construct a botanic garden, but his plans were eventually put aside. In 1621, Worm personally took over management of the neglected Garden and introduced a great number of Danish medicinal plants as well as rare foreign species he received from his many professional contacts abroad.

The Garden is the fourth in the succession of university gardens, which was established in 1872 on the former fortifications area of the city. The rock gardens and other higher areas are part of the old ramparts, while the lake is a remnant of the old city moat. A fully functional garden first became possible in 1769, when King Christian VII donated 2,500 Daler to the University, the interest from which would be used for the Garden's clearing and cultivation. The next year saw the King grant a part of Oeder's botanical garden near Amalienborg Palace to the University, as it became clear that plans to expand the Old Garden would not be realised. Christen Friis Rottboell, 1727–1797, a professor of medicine, managed to erect a greenhouse before Oeder's botanic garden finally closed down during 1778.

Throughout the 24 acres of the Garden there are many interesting features and collections. Particularly noteworthy is the Palm House, with tropical and subtropical plants, orchids, cacti and other succulents, and a historical collection of cycads. Other greenhouses display collections of orchids from Thailand, plants from Madagascar, bromeliads and insectivorous plants. In the outdoor section one can find arctic and alpine plants, wild Danish plants, perennials, annuals (probably the largest collection in the world), and tuberous species such as Cyclamen, Crocus and Fritillaria, among countless other varieties of flora.

In the Botanic Garden of Copenhagen, 25,000 living specimens represent a total of more than 13,000 species. In comparison, the total flora of Europe numbers some 11,000 species, and the total world flora of higher plants is estimated to be between 260,000 and 300,000 species.

The mission of the Garden is to maintain a rich collection of living plants for research, education and general information purposes and thereby contribute to increasing the knowledge about plants and to promote an interest in nature and it's wealth of resources, both nationally and globally. Accordingly, the objectives and tasks of the Garden are to maintain a rich and diverse collection of living plants, outdoors, in greenhouses, and in a genebank, for use in research and education, maintaining the collections through regular acquisitions. It also plans to foster understanding and awareness of plants and to encourage a general interest in nature and natural resources through public displays and conduct research, primarily in topics related to the living collections.

The botanical gardens of the world are key institutions in botanical research and contribute to the conservation of rare, threatened and potentially useful plants (as for example wild relatives of crops). Thus biological diversity and cell and tissue culture are headlines for the research at the Botanic Garden of Copenhagen University. To facilitate the research, the Garden has recently established a modern cell and tissue culture laboratory and modest but modern genebank facilities.

The Garden currently holds 22,052 living plants which all are registered in the electronic database, and which represent 359 families, 2,984 genera and 13,210 species. In addition, a large number of subspecific taxa and experimental plants are not yet registered in the database, but represent a significant part of the Garden's collections. The genebank holds seeds of about 800 wild Danish species, some 1,000 species from the Garden's international collections, and about 150 accessions of seeds classified as particularly relevant for research projects. More than 50 species are maintained through tissue culture in the laboratory, and among these are several species that are critically endangered in the wild, such as varieties from the Galápagos Islands in the Pacific and the Mascarenes in the Indian Ocean.

Plants which are primarily cultivated for information purposes are placed and grouped accordingly, as for example in taxonomic groups (the orchid collection and the Sorbus collection), in biological groups (the annual section, and the carnivorous plants exhibition), in ecological groupings (the swamp meadow and steppe sections), in geographical groups (the Danish plant section, or the Madagascar collection) or in climatic groupings (the Palmhouse and the Succulent House). These designated plant beds offer the visitor a means from which to understand the different parts of Copenhagen's collection according to various scientific lines in enquiry, as well as present a pleasant environment in which to observe the natural splendour of the Garden's collections.

tag: Paintbrush Lily or Snake Lily (*Scadoxus puniceus*) is a South African plant belonging to the Amaryllis family. photograph: The Botanic Garden. previous pages top left: The Weeping Willow (*Salix babylonica*) is one of the first trees which were planted in the garden in 1873. photograph: The Botanic Garden. previous pages bottom left: The bridge over the Botanical Garden's lake. The lake is a section of the historical fortifications that surround Copenhagen. photograph: The Botanic Garden. right: View across the lake in the Botanic Garden. In the background is the Palm House. photograph: Ole Hamann. left: The Botanic Garden's lake in autumn. photograph: Ole Hamann.

ESTONIA
TALLINN BOTANIC GARDEN

The terrain on which the Tallinn Botanic Garden is situated is fundamental to its plant collection. The Garden is located on the eastern outskirts of the Estonian capital, very near to the coastal district of Pirita. The temperate climate of this area is due to the moderating effect of the Baltic Sea, which has greatly influenced the success of certain species since the Garden's establishment in 1961. With a little help from warm greenhouses, Talinn cultivates flora from all over the world and within a quick stroll visitors are transported from European meadows to African tundra, or from desert to rainforest.

Tallinn Botanic Garden has an intrinsically rural essence, not shared with most other botanic gardens. Rambling, mountainous and undulating, the landscape here is dramatic and varied, divided by the long sweep of the Pirita River. The relief is predominantly formed by ancient coastal dunes, and the altitude reaches 24 metres in some places. Some of the original vegetation of the site is still visible in many parts of the Garden, lending it a more organic feel than many of the world's more architectural gardens. In fact, almost 500 vascular plants growing here have been listed as native to this exact spot in Kloostrimetsa. These are mostly classified as heath and nemoral forest species.

The curators at Tallinn correspond and exchange frequently with their neighbours in efforts to accentuate, first and foremost, this comprehensively vernacular collection. As part of the Soviet Union, the botanists here would frequently organise expeditions around the region, from Arctic Siberia to Transcaucasia to Central Asia, in attempts to create an anthology that was inherently geographical, inherently Soviet and, by extension during this period, surreptitiously political. Still now, the Tallinn botanists frequently engage with the Russian Academy of Sciences Botanic Garden in Moscow, as well as with Latvian and Lithuanian botanical institutions.

Its focus has always been international however, and the Garden is also home to a multitude of global species, including plants native to Australia, Madagascar and the Canary Islands. Exotic plants such as the Date Palm, Cocoa Tree, and Coffee Tree are all cultivated here, housed in the expansive greenhouses that covers an area of 2,100 square metres.

Most impressive about the Garden is its commitment to civil education. Regular tours are led around the grounds, including a nocturnal walk through the scented gardens in the summertime. There is also an extensive exhibition schedule in which a new range of exhibits—be it poisonous plants, mushrooms, or culinary plants—is shown almost every month in the Palm House.

tag: Koit, a type of rose that is cultivated locally. opposite: The The Palm House. below: One of the seven artificial ponds in the Botanic Garden. right: A variety of succulent plants in the Garden's Greenhouse. photograph: Tarmo Niitla.

UNIVERSITY OF OULU BOTANIC GARDENS

Originally established in 1960, the University of Oulu Botanic Gardens were relocated from the city centre site in the summer of 1983. The Gardens are now located on the northern edge of the Linnanmaa University campus, close to Lake Kuivasjärvi, and constitute an integral part of the Department of Biology at the University; their main purpose being to provide living plant material, experimental plots and practical help for botanical teaching and research. They also serve as an important educational resource for schools and a recreation area for the public.

The Gardens now cover nearly 40 acres, and two-thirds of that is devoted to open-air paths and gardens. The site has undergone significant artificial landscaping, developing hillocks and rockeries to suit a variety of vegetation as well as small-scale lakes and water features. The outdoor gardens are designed to a classical constraint, divided into seven sections showcasing over 4,000 species of plant life. The collection is arranged systematically according to plant type and natural habitat, showcasing the mutual relationships between similar plants from varied origins. Medicinal plants and those with an economic crop value, such as food and industrial use warrant their own sections, as do plants falling under the category of Fennoscandia, that is, plants native to the Finnish and Scandinavian landscape. These include a special selection of ferns, mosses and lichens, and these native specimens are bedded in artificially created meadow and shore habitats to evoke their natural environment. The hilly areas contain a rock garden, which benefits the plants native to fjells, mountains and other specifically north European terrain; whilst mountainous plants from Asia, North America and other European regions are displayed on separate hills.

The section devoted to ornamental plants has been planned with a view to stimulate interest in growing and maintaining private ornamental gardens; and much of the research undertaken at the Gardens is in the experimental cultivation of plants of different provenances, to test their suitability for ornamental purposes in northern Finland.

The Gardens, which are among the northernmost in the world, have an important role in supporting research on northern plant types and the suitability of plants for a northern climate; consequently many of the specimens exchanged move between gardens in Russia, Canada, North America and Northern Europe. However within its limited area, the Gardens do host a significant collection of plants from warmer climes in their two unusual, and state of the art, greenhouses. The greenhouses were designed by the office of architect Kari Virta, as were the buildings belonging to the University; consequently there is a harmony of design ideology across the institution. However, the structures are noteworthy in their own right; constructed from steel frames covered with cellular plastic, they have an unusual pyramid form which allows for plants of diverse heights and shapes to grow without overshadowing one another and obstructing the display. The two pyramid shaped greenhouses are affectionately known as Romeo and Juliet; the smaller of the two, Juliet, houses Mediterranean, temperate and succulent plants, while the taller, Romeo, showcases a selection of tropical plants and lianas, as well as a selection of tropical economic crop trees. A connecting hall contains carnivorous plants and orchids, as well as housing a mini pyramid principally dedicated to ferns.

A large woodland arcs round the eastern face of the outdoor section, completing the spectrum of plant life to be enjoyed at the Gardens. The Arboretum is maintained in a naturalistic style, with trees planted in geographical groups; several winding paths guide the viewer through areas of natural beauty and culminate on the southern shores of Lake Kuivasjärvi, presenting breathtaking vistas.

tag: *Phlaeonopsis,* part of the Subtropical Greenhouse Collection. opposite top left: View of the Native Mountain Collection in winter. opposite top right: The Ornamental Plant Section. opposite middle right: An Orchid (*Coelogyne cristata*), part of the Subtropical Greenhouse Collection. opposite bottom: The Subtropical Greenhouse at night. all photograph: Collection of Botanic Gardens.

FRANCE
BOTANIC GARDENS OF MONTPELLIER

The Botanic Gardens of Montpellier were born from the drive to medical research in the twelfth century. Montpellier, crowned with towers and church steeples, was a sea-trading city as well as a fortified cloister. The ships of the *speciadors sovereigns* (rich bourgeois citizens) continually brought spices, medicines and plant specimens from all over the world to the doors of the medical university situated in Montpellier. During the sixteenth century the university built a hall for dissection along with a small garden to develop their already well-respected medical botany research.

The Botanic Gardens of Montpellier as we know them were established by King Henry IV in 1593, with the help of botanical pioneer, Pierre Richer de Belleval. The vegetables and medical plants (or simples) were placed with great ingenuity in a variety of environments, ranging from sands, rocky grounds, wet soil and a curious aquatic labyrinth. The simples collection attracted interest from scholars the world over, with visitors from Germany, Switzerland and England travelling to marvel at the success of France's oldest botanical gardens.

During the nineteenth century, the historical gardens underwent significant expansion, more than doubling in size. A succession of well-respected directors bolstered the Gardens' reputation, as well as accomplished botanic artists, including Touissant Node-Veran, who catalogued a significant amount of plants and mushrooms for the scientific archives, culminating in 972 vellums and 679 drawings.

The two World Wars were difficult times for the Botanic Gardens of Montpellier, and the grounds suffered much damage through warfare and neglect during this period. Between 1950 and 1960, landscape gardener Herve Harant undertook a wholesale renovation of the Gardens, returning the grounds to their former glory. Continuing to delight and surprise, the Gardens now offer several attractions including tropical greenhouses, a nineteenth century orangery, a systematics school and a gorgeous English garden. The 11 acres of grounds serve to impart valuable botanical knowledge to nearby universities while remaining a living chapter in the botanical history of France. Today, the Botanic Gardens of Montpellier remains one of the cradles of modern botany, and contributes invaluable research to the international scientific community.

tag and below: A wildflower meadow leads up to the Gardens' display of towering palms. opposite top: The Gardens' Orangerie. opposite bottom: The Botanic Gardens of Montpellier are the oldest botanical garden in France; here a systematic garden carries on the tradition of scientific and academic research.

BOTANIC GARDENS, PARIS

Following on from the tradition of the Royal Medicinal Gardens and under the instruction of King Louis XIII, Botanic Gardens opened to the public in 1640 and is the principle, as well as the oldest, botanical garden in Paris. The Garden contains the Musée National d'Histoire Naturelle within its grounds, which it shares with several other museums and sites of interest, including the erstwhile Royal Menagerie or zoo, the Grande Galerie d'Evolution, the Galeries de Paléontologie et Anatomie Comparée and the Galerie de Minéralogie et Géologie.

Flowerbeds and buildings conjure up the period during which they were conceived, and many of the features of the Botanic Gardens reveal the social history of the grounds. The Gardens were originally planted by Guy de La Brosse, Louis XIII's physician, in 1626 as a private garden for the King, known as the Jardin du Roi. The next century saw the Botanic Gardens pass through several phases of expansion and diversification; 1739 saw the addition of a labyrinth, whilst in 1792 the Royal Menagerie was shifted from Versailles to the Paris grounds. The Mexican Glasshouse, built by Rohault de Fleury from 1834 to 1836, is an early example of cast iron architecture in France and precedes the more famous realisations of Gustave Eiffel by a century, as does the Gardens' metallic Buffon Gazebo and ampitheatre. Made entirely from glass and metal, it is in fact the largest greenhouse in the world, housing a fine collection of succulents and arid plants. The Wintergarden Greenhouse is a stunning example of Art

Deco architecture and provides a lush encasement for the tropical plants, the graceful curves of the high domed structure echoing the lianas and exuberant forms of foliage within. It is not only the architecture that holds historical significance; the oldest historical tree in the Botanic Gardens is now 366 years old, and indeed the entire Gardens and the interior of the surrounding walls are officially listed as historical monuments.

Designed and maintained in a formal style, the Gardens contain several outdoor features conceived to enchant and intrigue the viewer; curiosities such as the labyrinth hillock, made from refuse and chalk rubble at the beginning of the fourteenth century, or the classically delightful Allées Buffon, lined with London plane trees making for a romantic promenade. Between the two paths are the five *carrés de la perspective* or squares of perspective, symmetrical rectangular beds which draw the eye across the whole of the Botanic Gardens, flowering from April until the end of autumn.

The Gardens feature a full and fragrant rose garden that runs the length of the Galerie de Minéralogie, and is dedicated exclusively to this 'queen of flowers'. It was designed and planted in 1990, with its main theme being the history of the rose species, showcasing many old and delicately scented varieties as well as modern breeds and hybrids. The central alley, shaded by climbing roses, is hemmed by beds inviting the visitor to discover and smell myriad varieties of

the flower, and the gardeners hold annual free training sessions on the pruning of roses for the general public.

The Alpine Garden, nestled at the heart of the site, boasts a collection of over 2,000 species, gathering together plants originating from French mountain ranges as well as other around the world. These fragile plants are carefully tended and monitored to allow the visitor the rare privilege of glimpsing the various flowering seasons. A more homely, though no less pleasing, sight is offered by the Perennial Garden, where over 450 species of hardy perennial plants are comfortably arranged in brick lined beds around a centrepiece of 150 varieties of irises.

For all its stunning architecture and formally designed features, the Gardens are foremost a place of learning and enlightenment. The first botanical school on site has been in existence since 1635, and was completely renovated in 1954. The Gardens enable applied research into phenology and pollens to be undertaken. However, for the visitor it is the impact of the ground's luscious design, rich content and historical legacy that shape the experience of the Botanic Gardens.

tag: Spring flowers in the Gardens. opposite page: View of the Gardens from the Cabinet d'Histoire Naturelle. above: View of the Greenhouse.

GERMANY
MUNICH-NYMPHENBURG BOTANIC GARDEN

Established in 1812, the Munich-Nymphenburg Botanic Garden was originally situated in Munich's city centre; however, due to recurring cession of territory towards the end of the nineteenth century, it was no longer of a functional size. Therefore in 1914 a new site of 54 acres was obtained on the fringe of the city. Since 1966 the Garden joined with the neighbouring Botanical State Collection and the Institute for Systematic Botany of Munich University to create a single facility for the study of botany and horticulture. With the State Scientific Collection and the adjoining museums, the Garden now comes under one administrative body, whilst the Nymphenburg Schlosspark to the south afforded perfect protection to the newly laid out Garden from any future spatial developments.

The glasshouses and the outside garden currently harbour approximately 14,000 different plant species from around the

world. Amongst its outdoor exhibits, the Botanic Garden features the Schmuckhof, which is a sunken formal garden laid out in front of the Institute buildings, opening to a cafe on the south face. The Schmuckhof contains a waterlily pond in its dry stone walled enclosure, which is planted up with dwarf shrubs, ornamental beds and a noteworthy collection of Tree Peonies. A highlight of the Schmuckhof's rotational seasonal planting is a luxuriant display of dahlias and asters in late summer; concurrent with themed exhibitions displayed in the Winterhalle or Orangery, and these are preceded in the spring months by a colourful selection of tulips along the Garden's ample lawns. A small, prettily landscaped path called the Spring Garden also features plants that come into blossom early in the year; and this area runs parallel to the Alpine Show House, which displays a selection of Mediterranean Alpine plants through panoramic windows. Peony and iris beds follow the

springtime alpine display, as do over 200 species of May-flowering rhododendrons, planted in a naturalistic fern glen to the south and west of the Garden.

The Arboretum takes up the west side of the site, hosting mighty beech trees as well as giant-flowering magnolias; once again this area is naturalistically planted and left unmown, as is the Bavarian Plant Community area, in which a wild mesophytic forest native to the environs of Munich, is encouraged. Contrastingly, the Medicinal, Economic and Systematic Order Beds are meticulously laid out, the latter displaying over 1,600 species in concentric formation.

The Garden maintains several glasshouses with specific functions; from the sides of three large connected glasshouses lead off various smaller glasshouses containing more specialised collections.

The first of the large glasshouses displays plants from arid regions of the New World, while the third of the large glasshouses contains arid plants from the Old World. Between them lies the Tropical House with an impressive cupola rising to 20 metres. This amply provides the necessary height for the tall bamboos and high canopy plants exhibited here in a hot and humid climate. The smaller Orchid House shelters a rich and varied display of blossoms, chosen from the 2,000 orchid species the Garden has in reserve. The Tropical Economic Plant House displays examples of useful cash crops for food and industrial purposes. Other smaller glasshouses include the Water Plant House, Succulent and Cacti Houses and the multipurpose Victoria House, and still more in a separate constellation of structures; a Tree Fern House, Cycad House and one for bromelias are followed lastly by a Temperate House, home to pot plants with an attached glasshouse for epiphytic ferns.

With its expansive and comprehensive collection, wide selection of environments and high level of technological maintenance, the Munich-Nymphenburg Botanic Garden is one of the most important botanic gardens in Germany.

tag: Pathway leading to one of the Garden's many Greenhouses. opposite: The Garden's Succulent Collection, which can be found in the Mexico House. above: The Waterlily House. below: Central flower beds in front of the Garden's Schmuckhof.

GERMANY
BERLIN-DAHLEM BOTANIC GARDEN AND MUSEUM

The history of the Berlin-Dahlem Botanic Garden and Museum is just tumultuous and as extensive as the history of the city itself. Its first ancestor was a small kitchen garden in the city, which was then expanded in 1679 when instructions were given by the Grand Duke to open an agricultural model garden. A site was duly chosen in a suburb of Berlin called Schöneberg. At the time, Schöneberg was a small suburb on the very edges of the city, though it was not long before the enveloping jaws of Berlin progressively ate up the village as it poured outwards. The Garden was further developed with the aim of propagating and conserving Germanic flora and exotic plants collected from German colonies.

The main tropical greenhouse at Berlin—located at its new premises in Dahlem—is one of the largest in the world, and is only one of the 16 constructed here. Palatial, gleaming, and Gothic, the greenhouse features rock pools and waterfalls, as well as the famous two-leaved *Welwitschia* plant native to the Namibian desert. The other greenhouses are divided into geographical sectors and climate zones, providing a loose survey of the planet's vegetation. Highlights include the candelabra-shaped Euphorbia plants in the Old World Garden, and the bed-sized waterlilies in the Victorian House.

The Garden now spread well over 120 acres, making Berlin's one of the world's largest botanic gardens. In dealing with such a vast site, the curators have long focused on thinking beyond the usual geographical survey of displaying plants, and have come up with some incredibly innovative ideas. In 1984, the Fragrance and Touch Garden was opened, created specifically for disabled visitors. The focus here is on the sensory experience—smell and texture rather than sight—and the plants have all been chosen for their appeal to the nose or the hands. There is also the Medicinal Garden, where 230 medicinal species are arranged in the shape of the human body, complete with explanatory signage.

Besides the Garden, there is also the Botanical Museum. Including sections devoted to palaeontology and phytogeography, the Museum contains numerous fossils and plant formations reconstructed in the form of miniature, artificial landscapes. It also houses a department devoted to Ancient Egyptian horticulture, in which can be found botanical relics and reconstructions of flower garlands and other herbal decorations from the tombs of the pharaohs.

tag: *Guzmania sanguinea,* which can be found in one of the Warm Greenhouses. below: A historical map of the Garden. opposite top: The Cacti Collection, which can be found in one of the Cold Greenhouses. opposite bottom left: The Mediterranean Greenhouse. opposite bottom right: The Italian Garden, with the Main Tropical Greenhouse in the background.

A burning ambition of the garden architect and botanical trader, Heinrich Siesmayer, coupled with the opportunity to purchase a collection of exotic flora from Duke Adolph of Nassau, led to the foundation of the Palmengarden in Frankfurt in 1868.

Apart from the traditional *Palmenhaus* building (which is still used for swanky society functions, concerts and exhibitions) the Garden has changed dramatically in recent years to provide the twenty-first century visitor with a colourful selection of flora and fauna. Devotedly tended rhododendron, heather, rose, shrub, woodland and rock gardens, together with an area dedicated to the new German style drift garden, provide outdoor interest. The multitude of tulips and crocuses in the spring, along with narcissi and hyacinths give way to azaleas and rhododendrons in May and June. Roses, particularly in the new Rose Garden, are in full bloom at the height of summer. Haus Rosenbrunn, situated at the centre of the Rose Garden, is garlanded by climbing blooms, in a spectacular display of some of the Garden's prized breeds. Day lilies, irises and peonies are displayed in broad herbaceous borders along the pond and the stream.

The Palmengarden is known worldwide for its ample collections of tropical plants. A spectacular selection of cacti and succulents on show offers a warm and welcoming contrast to visitors emerging from the Sub-Antarctica House. The complex of seven large modern greenhouses of 600 square metres each, together with six smaller glasshouses, form the newly constructed and computer-controlled Tropicarium, where visitors can experience not only the various climatic zones but also the diverse range of plant life from the rainforest to the desert. The Tropicarium is one of the largest complexes of greenhouses in the Garden. It displays tropical plants including orchids, bromelias, palms, ferns, succulents and many others species from all over the world. These plants are arranged mostly according to their natural habitats including specimens from the rainforest, mangrove, mountain rainforest, savannah and thorn forest regions. The large *Palmenhaus* was constructed in 1869 and built as an addition to the *Gesellschaftshaus*. It displays many palm species as well as a variety of other, mainly sub-tropical, plants. The new gallery of the *Palmenhaus* is used for flower shows and informative exhibitions and rare blooms are on display in the horticultural *Blütenhaus*.

tag: *Lathyrus vernus*. photograph: Klaus Lorbach. opposite top left: A woodland stream, surrounded by luscious plants. photograph: Klaus Lorbach. opposite top right: Pathway into the Garden's woodland area. photograph: Klaus Lorbach. opposite bottom left: Elaborate fungi growing on one of the Garden's trees. photograph: Manfred Wessel. opposite bottom right: View of the Herb Garden. photograph: Manfred Wessel.

GERMANY
HERRENHAUSEN GARDEN, HANOVER

The Herrenhausen Garden in Lower Saxony's capital of Hanover provide constant and unmistakable clues to its long regal history. This was the estate built and progressively developed over a period of centuries by the long line of the kings of Hanover. With the Herrenhausen Castle as the estate's centrepiece, their expansion was so accomplished that the Garden situated around the palace is split into four large, self-sufficient areas. The Great Garden is surely the most important and noble of them all.

The city of Hanover owes this jewel of baroque garden architecture to an extraordinary woman: Electress Sophia of the Palatinate. A doyenne of European Court life, she was the granddaughter of James I of England, the youngest daughter of Frederick V, Elector Palatine and, wife of Ernst August. The couple resided at Herrenhausen, and Sophia set herself the project of completely overhauling the surrounding grounds. Inspired by the Sun King, Louis XIV of France, she had this 150 acre garden laid out in the French style at the end of the seventeenth century, commissioning the French landscaper Martin Charbonnier to lead the design. Influenced by the formal Baroque gardens of France and Holland, Charbonnier created a geometric garden, divided into flower borders punctuated by statues of allegorical figures and Roman gods. Symmetry was key to Charbonnier's design—taking its lead from the seventeenth century love of collecting and organising

nature—and he implemented axial prospects, *parterres*, round basins and hedging to structure the space. Despite the rigorously strict approach to garden design, Sophia was keen on constructing an amusing recreational space: a site to entertain and receive members of European courts. A maze garden, open air theatre, cascades and numerous grottos all formed part of her masterplan to create the most supremely entertaining of outdoor spaces. Given her love for the grounds, there is a certain poetry to the fact that, during an evening walk in 1714, Electress Sophia actually fell to her death in her beloved garden.

Without Sophia's patronage, the Garden's further development proceeded sporadically. In 1846, almost 150 years after her death, work was begun in the *Berggarten* (originally meant as a vegetable garden, then made into an exotic garden in 1686). A conservatory was designed by architect Georg Ludwig Friedrich Laves, built for the purpose of containing palm trees. Within five years of its completion, the building housed the most valuable and extensive collection of palms in all of Europe. During this time, work was also begun by Laves on the Garden's mausoleum—Sophia's son, King Ernest Augustus I and his wife Queen Frederica were the first to be interred there. Despite these efforts, the vast majority of the *Berggarten*, and the Herrenhausen Castle, was destroyed by British air raids during the Second World War. The connections between the Hanoverian monarchy and the British Royal Family (Sophia was in line to the British throne, second only to Queen Anne; likewise, Sophia's son was King George I of England), were a matter of political contention when the Garden fell victim to British combat activity.

In some ways, the wartime destruction of vast sections of the Garden allowed important contemporary development that otherwise may not have happened. It created space and opportunity for the modernisation of an otherwise listed, unalterable landscape. A replacement for Lave's Palm House was erected, named the Rainforest House (*Regenwaldhaus*), now housing a tropical landscape that includes both palm trees and orchids, as well as different species of butterflies and birds.

Updating the nineteenth century concept of the steel and glass greenhouse, the construction is reminiscent of a gigantic leaf and was built as part of EXPO 2000, hosted by Hanover. Another exciting contemporary project exists in one of the historic grottos, where artist Niki de Saint Phalle transformed the arches by adding various items, including crystals, minerals, glass and seashells. Between 2001 and 2003, when the exhibition opened, de Saint Phalle and her co-workers covered the walls and interior with mosaics of moulded glass and mirrors. The space has taken on a palatial aura, and although comparisons to the Hall of Mirrors at Versailles may not be entirely accurate, it would make a neat link to Electress Sophia's original intentions for her garden.

tag: Plants in the Tropical Winter Garden. photograph: Nik Barlo Jr. previous pages left: Historical depiction of the Palm House, a conservatory designed by architect Georg Ludwig Friedrich Laves in 1846. photograph: Historic Museum of Hanover. previous pages right: The Schmuckhof, which links the Library and the Tropical, Cactus and Orchid Buildings. photograph: Nik Barlo Jr. above: Plan of the Herrenhausen Gardens. courtesy: Freunde der Herrenhäuser Gärten EV. left: The Desert Garden. photograph: Marita Heuchert. opposite: The Cactus Building, which features more than 1,000 cactus species. photograph: Hanover Tourism Service (HTS).

UNIVERSITY OF HEIDELBERG BOTANIC GARDEN AND HERBARIUM

The Botanic Garden of Heidelberg was founded in 1593 by Henricus Smetius primarily as a medicinal garden (*Hortus Medicus*). Despite this being a concept religiously employed by many universities during the sixteenth and seventeenth centuries, it was not until the mid-nineteenth century that the Garden was first administered by the Department of Botany at the University of Heidelberg. During this period, the Garden underwent several relocations, reopening in 1915 at its present site in the Neuenheimerfeld. The establishment of the Garden underwent another setback during the Second World War, when the entire collection was lost due to combat activities.

Bearing this in mind, it is remarkable to consider how the living collection has already built up to over 14,000 species in the short space of time between the Second World War and the Garden's present day inception. This is mostly down to the tireless efforts of professor Dr Werner Rauh, who was the director of the Garden from 1960 to 1982. During this period—and up until his death in 2000—Rauh undertook countless expeditions to the tropical, subtropical and arid regions of the Americas, Asia, and Africa, bringing back with him a vast selection of succulents, orchids and bromeliads. Rauh discovered and described many new species of plants on these adventures, and a great deal of flora at Heidelberg

are named after him, such as the *Amaryllis genus Rauhia*, the *cactus Rauhocereus* from Peru, and the Brazilian *orchid Rauhiella*. His contribution exceeded 10,000 species, most of which are now housed in the greenhouse complex. Furthermore, this collection at Heidelberg now holds an important collection of extinct and endangered species from around the world. Many of these plants' natural habitats have been subject to a progressive sequence of destruction, and they have become all but obsolete in their tropical homeland. The Universtiy of Heidelberg Botanic Garden and Herbarium has been an important conservation reserve in that respect. Particular conservation programmes in progress at the site involve, amongst others, the following species: the Blue Bellflower from Mauritius (*Nesocodon mauritianum*), Brighamia insignis (*Lobeliaceae*), a stem-succulent from Hawaii, the Cochlearia macrorrhiza (*Brassicaceae*), and endemic species from Eastern Austria.

The rare and tropical nature of these plant varieties have slowly become an important background to modern botanical research in the twenty-first century, including molecular biology, plant evolution, and pharmaceutical botany and ecology. The University of Heidelberg has witnessed increasing public demand for education in these biological, ecological and botanical sectors, and the University's Botanic Garden and Herbarium have proposed an expanded mission in response to this demand.

tag: A Purple Foxglove (*Digitalis purpurea*). opposite left: The Tropical Greenhouse. opposite top right: An example of the Aloe plant family, which can be found in the Old World Succulents Collection. opposite bottom right: The Alcohol Collection where thousands of extracts and tinctures are carefully preserved for medicinal value analysis and scientific use. left: A preserved example of an Exotic Bucket Orchid, which produces a special glue that enables pollination by bees.

GIBRALTAR
GIBRALTAR BOTANIC GARDENS

Gibraltar's military history meant that, for centuries, little thought was given to the improvement of the social or cultural life of its civilian inhabitants. In 1815, General Don was moved to establish a walk around the Grand Parade. The promenade around the parade was gradually expanded to include approximately 20 acres of land to form what is known as an Alameda, where the inhabitants might enjoy the outdoors protected from the extreme heat of the sun. These have become known as the Alameda Gardens, derived from the Spanish word *Alamo*, or White Poplar (*Populus alba*); indeed old records mention these shady trees growing along the Grand Parade. The walks opened to the public on 14 April 1816.

The Gardens were laid out with numerous interconnecting paths and terraced beds; improvements through the early years included the introduction of gas lighting and the erection, possibly in 1842, of an archway made out of the jaws of a whale. The Gardens reflect much of the military history of the island; indeed, the Grand Parade itself was the location for the weekly changing of the guard

amongst other ceremonial occasions typical of Gibraltar's martial past. In order to reflect the significance of the military presence, several monuments and relics remain; to this day two guns on slides overlook the Grand Parade from the east.

A colossal statue of General Eliot, now relocated, was carved from the bowsprit of a Spanish Man-O-War and resided here until it was replaced in 1858 with a bronze bust, surrounded by a proud display of four Howitzer firearms from the 1780s. Three years after the opening of the Alameda, another bust of the Duke of Wellington was ceremoniously unveiled. It stands impressively on a marble pillar brought from the Roman ruins of Lepida or Libya. Around Wellington's column stand two 33 centimetre mortars and a 1758 bronze gun on a wooden garrison carriage.

The plants of the Gardens are a combination of native species and others brought in from abroad, often from former British territories like Australia and South Africa with which Gibraltar had maritime

links at the time of the British Empire. Several trees, such as the Ancient Dragon, Wild Olive, and Nettle Tree thrive here, whilst the Stone Pine contributes its delicious nuts to the local delicacy of sugared, roasted *piñones*. The Gardens also boast an impressive collection of succulents, and several varieties of palm; which, though of many different origins, are happily suited to the warm and temperate Mediterranean climate. Vibrant floral blossoms include hibiscus, bougainvillea, geraniums, various daisies and lilies. Many of the finest examples of these flowers can be found in the Dell, an Italian-style ornamental garden laid out by a Genoese gardener in 1842. This picturesque area showcases many of the finest examples of local flowers, plants and trees, as well as several tropical ferns. These are planted around two early twentieth century fountains, a waterfall and a pond, which houses goldfish, frogs and terrapins. The Alameda also features an open air theatre, incorporating not only a multipurpose performance space but also a waterfall and lake—the largest area of open fresh water on the Rock, populated by Koi Carp and a collection of exotic lilies.

A seventeenth century stone cottage, once the head gardener's residence, rounds up the Alameda with an informative display on the botany and natural history of Gibraltar and the Gardens in particular, including the history of the Alameda. The Gardens' ecological concerns include the conservation and establishment of a living collection representative of Gibraltar and its hinterland and the display and dissemination of information regarding its exotic collection.

tag: The Gibraltar Candytuft (*Iberis gibraltarica*), growing in one of the Native Plant Areas. This species grows wild only in Gibraltar and North Morocco. photograph: Rock Interactive Ltd. opposite left: Postcard of the The Sunken Italian Garden, known as The Dell, in 1910. opposite right: A Mexican Fan Palm (*Washingtonia robusta*) in The Dell. photograph: Rock Interactive Ltd. above: A Red Admiral Butterfly on *Ceanothus arboreus* in the California Bed. photograph: J Cortes middle: Stone Pine woodland and the Australian Bed off the Garden's upper Main Walk. photograph: Rock Interactive Ltd. below: The Italian Garden in The Dell, with the Upper Rock Nature Reserve in the background. photograph: Rock Interactive Ltd.

GREECE
BALKAN BOTANIC GARDEN AT KROUSSIA MOUNTAINS

Greece is a country of great botanic wealth. Due to its geographic position and the coexistence of three floral regions, the plant-life of Greece—in proportion to its area—is one of the richest in Europe, consisting of approximately 5,700 species of higher plants.

The dangers to plant biodiversity in Greece are well-known. Fire, land reclamation, overgrazing, and ever-increasing human intervention have long disturbed the integrity, stability, and beauty of natural Greek plant communities. The need for landscape protection and the conservation of biodiversity is becoming more pressing.

The Balkan Botanic Garden at Kroussia Mountains is one of the youngest botanic gardens in Europe. It is a new scientific institution in Greece dedicated to conservation, botanical and horticultural research and environmental education. Its mission is to grow, study and protect the Balkan Flora in parallel with the development of ecotourism, providing an aesthetic, educational and recreational resource for the Balkans and for international visitors. In addition, it employs scientific, technical and manual human resources. The Garden occupies an area of 74 acres and covers the length of a hill ridge on the Kroussia Mountains located in Central Macedonia, Greece, near the village of Pontokerasia, about 70 kilometres north of the city of Thessaloniki.

Since its conception in 1997 major works have been completed on the Garden, such as the construction of an exhibition and office building at Podokerasia and the building of the Horticulture Research Institute on the campus of the National Agricultural Research Foundation (NAGREF) in Thessaloniki. 5,600 metres of metal fencing was erected on the site, roads and pathways were constructed, two artificial lakes connected by a brook fed by rainfall were excavated, and irrigation systems were installed. These developments served to enhance the visual qualities of the Garden, while providing important irrigation and water feed to the collection.

The natural vegetation, which does not differ significantly from that outside the Garden, features about 37 acres of trees, mainly oaks, such as *Quercus pubescens*, *Quercus frainetto* and *Carpinus orientalis* along with a wealth of herbaceous perennials. The remaining acreage has been cleared for new plantings of flora found throughout the Balkans.

The various sections of the Garden are arranged according to different themes. These may be taxonomic relationships, ecological adaptation or different life histories. The scope is mostly educational, however, the display of the living plant collection attempts to convey an attractive setting that enables people to enjoy and appreciate the recreational and horticultural value of a botanic garden. The landscaping of the two and a half acres surrounding the exhibition hall attempts to combine traditional Greek garden elements (genius loci) with the demands of a modern botanic garden. It has been landscaped to show the diversity that exists in useful plants, to create a local garden with a sense of place, to demonstrate low input gardening and finally to show an unobstructed view toward the mountain range and the main body of the Garden's collections.

A selection of traditional and old Greek fruit tree cultivars were planted in the far corner of the building site, in accordance with the Greek tradition to concentrate 'useful plants' around their houses and in their gardens. Aromatic and pharmaceutical plants are given much prominence, given that Greece is one of the centres of diversity of the *Labiatae* family, members of which grow in the Garden on a steep slope that provides the best conditions (a sunny, dry, stony environment) for oregano, sage and mint

An authentic representation of each of Greece's five (often overlapping) vegetation zones, each of which has a distinct ecology, character, flora and history is displayed within the main Garden grounds. They include a Mediterranean evergreen forest zone, sub-Mediterranean, Beech-Fir forest zone, montane

coniferous forest and an above tree-line zone. Rocks and crevices in an artificial alpine environment display the neat growth habit and bright colours of alpine plants. Another section consists of maquis and phrygana; short shrubs evolved from overgrazing and logging, which adapted to withstand these man-made conditions by developing cushion shapes and thorns. A spring and an autumn meadow display Greece's wide diversity of bulbs and geophytes appearing in each of these seasons. Also, there are plans for an arboretum of about 375 woody species native to the Balkan Peninsula, arranged according to their phylogenetic relationships.

A small path leads off to a hidden natural pond, the setting for a fern and Campanula collection, surrounded by rocky outcrops and remnants of an old stone road. Part of the Garden is set aside as a fallow field for wheat and the annual weeds that used to be common in agrarian fields but have been dramatically reduced by herbicides. A comprehensive collection of wetland plants around the two artificial lakes completes the overall image of the Garden.

tag: Orchid (*Cephalanthera longifolia*). above: View from the Oak Forest overlooking the Garden's lake. top right: A section of the Water-loving Plants area, comrising 1,750 plants belonging to 50 species. bottom right: View of the Aromatic and Medicinal Plants Section of the Garden that includes 3,500 plants belonging to 26 species. all photographs: Balkan Botanic Garden at Kroussia Mountains.

In relation to the size of the country, Greek flora is among the richest in the world. There are approximately 5,700 species and sub-species of plants, of which approximately 1,150 are indigenous. This number will increase during the following years, as new species and sub-species are being discovered in far removed places.

The plants of Greece can be classified into three basic groups those that grow in alpine and sub-alpine zones, at an altitude of above 1,800 metres; those of the middle mountain zone, at an altitude from 700 metres to 1,800 metres; and the Mediterranean plants which grow next to the shoreline, on the plains and the shrub areas and forests of the lower mountain zone, at an altitude below 700 metres.

The Philodassiki Botanic Garden includes mainly Mediterranean Greek flora, as with approximately 2,000 species and sub-species it accounts to roughly a third of the total of Greek flora. Obviously, the varieties mostly represented are those of Mount Hymettus, where the Garden is located, and of the Attiki region in general. It also displays many elements of the Mediterranean ecosystems: it has a rich paeonia vegetation, consisting of large bushes and certain coniferous vegetation, mostly evergreens. Another important element of the Garden is its brush bushes, which are characteristic of the dry bush vegetation of the Mediterranean landscape. In these seemingly barren spaces, a great number of flora grow, including perennials, annuals and bulbs, giving a brief but striking floriferous look to the Garden, mainly during spring months. In addition certain sites, with deeper and more humid soil, such as the ones close to the small stream which crosses the Garden, encourage the growth of mountain plants that require increased humidity to flourish.

The Garden was established in 1947 on a site offered by the adjoining monastery. Initially 200 species of plants had been collected. As the years went by, the priorities of the Garden changed. Mainly due to a lack of funds, the grounds turned 'wild' in a very specific way, governed by the judicious eyes of the gardeners, resulting in a very particularly Greek charm and natural character.

Due to the existing structure of the Garden, the plants are displayed mostly in an ecological structure, according to their natural demands, such as soil, light and orientation: the cacti (*chasmophytes*) flourish in between rocks, the brush bushes within rocky, sunny places, and the more mountain varieties and plain growing plants in richer and irrigated sites. Wherever possible, there are groupings of one genus' species alongside another, such as the digitalis, the peonies or centauries.

The basic objective of the Garden is to educate, attempting to present as many varieties of Greek flora to the visitor as possible. Conservation is also an important aim for Philodassiki, with a focus on preserving endangered species whose biotypes have been destroyed for various reasons. Therefore, one of the objectives is to consciously make sure that all the species of flora do not become extinguished in the same way that the *Centaurean Tuntasia* and *Centaurea sibthorpii* have. Despite its rigorous policy of conservation, the Philodassiki Botanic Garden offers a unique and rambling experience of Greek flora that is as beneficial to the lay visitor as it is to the conservationist.

tag: Yellow Autumn crocus (*Sternbergia lutea*), a characteristic autumn flower of the Mediterranean and North Africa. opposite top left: Work in progress, which include cultivating the ponds with flowering Irises (*Iris pseudacorus*) and White Waterlily (*Nymphaea alba*). opposite top right: Spring anemones. opposite bottom: Entrance to the Botanic Garden.

BUDAPEST ZOO AND BOTANICAL GARDEN

The Budapest Zoo and Botanical Garden provides the exciting opportunity to experience and observe a truly vast range of flora and fauna from all over the world, within a lush setting. It is one of the oldest zoos in the world, inaugurated in 1866, and has been operating simultaneously as a public botanical garden since the opening of its Palm House in 1912. Budapest's past persists through the ground's unique architectural ensemble largely built in the Secessionist or Art Nouveau style of the early twentieth century. This is exemplified by the whimsical and imposing arched entrance gate resting on enormous sculpted elephants.

During the second half of the nineteenth century, the early establishment suffered a temporary lull due to various blights on animals and plants, lack of funding and a brief waning of interest. At that point the management resorted to showmen, raffles and comics to retain audiences, but the true calling of the institution was resurrected with the careful introduction of many specialised enclosures and gardens, under the stewardship of Károly Serák in the 1870s, who ushered in a fondly remembered golden era. Unfortunately, funding and management problems persisted briefly after this phase, which resulted in the city assuming control of the establishment. Famed Hungarian architects Károly Kós and Desző Zrumeczky were then employed to embellish the Zoo and accompanying gardens with early twentieth century style buildings, including the aforementioned main gate, the Monkey House, Palm House and many other modern structures. Since the 1990s, the Budapest Zoo and Botanical Garden has undergone major restoration and renovation, developing a state-of-the-art aquarium and a range of children's entertainment programmes.

Due to its rich collection of flora, the Garden is officially classified and protected as a conservation area. Presently it cultivates some 3,500 taxa, from which the display specimens are publicly exhibited. The collection harbours the common components of a botanical garden, including an arboretum of dendrology, with many examples of endemic and imported trees, and several climate-controlled environments housing displays such as the Cactus Succulent Collection, featuring plants adapted to arid

climates, and an Orchid Bromeliad Collection, with showy, gorgeous blossoms necessitating hot and humid conditions. As well as these conventional divisions, the Garden boasts curiosities like the National Bonsai Collection, the most prestigious collection of its kind in the country. There are specialised installations like the Japanese Garden, built in the 1960s over ruins from the Second World War, or the Rock Garden displaying a rich perennial flora and containing several local protected species. The joint zoological and botanical activity of this Garden provides an excellent base to present the variety and interrelation of Hungarian flora and fauna. The characteristics of the Garden's urban climate also enables open cultivation of several Mediterranean and subtropical species.

In order to ensure the continuous development of its collection, the Garden has long participated in the international seed exchange programme between botanical gardens of the world. The first seed catalogue was released by Garden in 1962 and current editions are delivered to more than 500 collection gardens around the world. The large number of visitors to the Zoo and Gardens enables it to play a prominent role in education and the raising of environmental

consciousness. Besides education and research, the Budapest Zoo and Botanical Garden also aims to represent gardening culture, by organising high-quality installations and horticultural events.

It was one of the first zoos worldwide to succeed in breeding giraffes, polar bears and Asiatic black bears, and to join preservation programmes for speces facing extinction, launched in the early 1920s. The Zoo can pride itself particularly on its success regarding the artificial insemination of the White Rhino, with the first calf born early in 2007.

Aside from the careful organisation of many educational and ornamental gardens, and of course a large collection of domesticated and exotic animals, a consistent striving for modernity and interest ensures that a visit to the Budapest Zoo and Botanical Gardens will be a highly rewarding experience.

tag: *Camellia Japonica*. opposite: View of a section of the Rock Garden. top left: Exterior view of the Garden's Palm House. above: The Elephant House. bottom left: Interior of the Garden's Palm House.

The Garden is a vast collection of resilient plant life that thrives in the dry and cold, dry climate of eastern Europe. It is the richest botanical garden of Hungary, with more than 12,000 species presented in a traditional landscaped garden. The hardy plants stretch around rocks and trees, poking through glistening frosted land around cool lakes.

The Garden was founded in 1827, inspired by English gardens of the Georgian age. In 1870 the grounds came into the possession of Count Sándor Vigyázó, devotee of science, who, together with Vilmos Jámbor, famous landscape gardener of his age, transformed it into a rich collection garden and arboreta. In the early years of the twentieth century the Garden was famous for the richness of its tree collection, rock gardens and greenhouses.

The Vigyázó family bequeathed their estate to the Hungarian Academy of Sciences, and the Garden became involved in the administration of the Academy of Sciences in 1952, after many mishaps. When the Research Institute for Botany was established here, the Garden began to grow exponentially. Damage caused during the countries war was repaired, and the ruins and bush cleared, revealing the original beauty of the landscape, enhanced by a much more varied collection of plants. In 1961, the Garden opened its doors to the public and since then has been an integral part of the Research Institute for Botany, a live laboratory and experimental area for researches.

The climate and soil conditions of the Garden are dry and somewhat hostile for plant life. The annual precipitation is barely more than 300 millimetres, similar to the dry areas of the Great Hungarian Plains, an area with an extreme climate, characterised by long summer droughts. The maximum temperature approaches 38 degrees Celsius in extreme cases, while the winter minimum can drop to as low as minus 30 degrees Celsius. There is scattered snow cover in winter, with early and late frosts and—in the low-lying areas of the Garden—fogs are frequent. The ground soil is calciferous wind-blown sand, bound sand and the matrix is solid clay.

The natural vegetation in the closer environs of the Botanical Garden is a magnificent representation of these extreme conditions. On the natural sandy lands at the end of the grounds, leather grasses flutter in the wind and the nearby hills are covered with oakwood. A fresh dash of colour is represented only by the narrow strips of riverside forests along the streams and by the bogs that dot the landscape. The remains of these can be observed along the garden lake and stream in the huge specimens of Hungarian ash (*Fraxinus angustifolia ssp. pannonica*) and the *Quercus robur.*

The extreme environmental factors preclude the growth of several plants associated with botanical gardens of more favourable climate, and the plant growth of the Garden can be maintained only by regular careful watering. At the same time, the visitor can see almost 1,000 characteristic species of Hungarian flora here, not to mention interesting plants from the Russian steppes, Central-Asian mountains, the Rocky Mountains in America, and trees and shrubs from eastern Asia.

tag: Wild Tree Peony (*Paeonia suffruticosa*). opposite left: A section of the Great Meadow in summer. opposite top right: The Rose Garden in summer. opposite bottom right: View of the Twin Pond with Caucasian wingnut (*Pterocarya fraxinifolia*) in the foreground. all photographs: Geza Kosa.

The Lystigarur Akureyrar Public Park and Botanic Garden has several aims, but the most important is to provide northern Iceland with trees, shrubs and perennials that fulfil the dual demands of beauty and hardiness required by the harsh landscape. The Garden also functions as a genebank for hardy plants well suited to Iceland's weather conditions. Stemming from this general idea, the Garden has myriad other functions, such as seed-exchange programs, education and recreation.

The Lystigarur Akureyrar Public Park and Botanic Garden is situated in the southern part of Akureyrar hillside. It lies on a moderate slope west of Eyrarlandsvegur road, and south of the Akureyrar Grammar School, reaching west to the school's pitch and south to Akureyrar Hospital. From its southernmost limit there is a view over the innermost point of Eyjafjordur firth, named Pollurinn. The oldest part of the park is in an almost unaltered natural state apart from two man-made grassy hollows.

Akureyrar Park was opened to the public in the spring 1912, becoming the first public park in Iceland. The aims of the Park Society were to make a park in Akureyrar, for the adornment of the town and as a place of recreation for its inhabitants. A botanic garden was founded in Akureyrar park in 1957. The Akureyrar Adornment Society recommended to the municipality of the town that they buy collections of living plants owned by

Jon Rognvaldsson, gardener and director of the park. As one of the world's most northerly botanic gardens, most of the species cultivated on the grounds have their origin in the Arctic, but the collection includes flora derived from temperate zones and mountainous areas.

Beds with species from Greenland, Norway and the Alps were established during the 1960s, with new beds for Icelandic flora developed in the autumn of 1985. The specimens were arranged ecologically rather than geographically in these new beds, with species that grow under similar conditions being grouped together. Here there are groups of species that grow or are common to different environs, for example the beach, herb-rich grassland, snow patches and in sheltered places. The newly renovated section for Icelandic flora currently houses 430 taxa, bringing the total amount of holdings to 7,000 species and sub-species of Icelandic flora, creating an important collection of rare Arctic plants and a special glimpse into their natural habitat.

tag: *Lewisias*. opposite top left: Autumn view of the tree-lined avenue, Vigdísarstígur. opposite top right: The Greenhouse, built in 1999, is a relatively new addition to the garden, enabling many more species to be preserved and viewed. opposite bottom left: A pond with a mirror-like surface which, were it not broken by projecting rocks, almost perfectly reflects this Betula. opposite bottom right: Perennial blossoms surrounding the cottage Eyrarlandsstofa. all photographs: Björgvin Steindórsson

INDONESIA
BOGOR BOTANICAL GARDENS

The fact that Bogor's secondary moniker happens to be the 'City of Rain' (*Kota Hujan*) is understandably rather beneficial for the botanical gardens situated in that city. 60 kilometres south of Indonesia's capital, Jakarta, Bogor was an important hill station during Dutch rule in the seventeenth century, and the city's position—halfway between the mountains and desert—made the spot perfectly conducive to housing Indonesia's foremost botanical garden. Lightning storms strike here regularly, apparently more than any other place on earth.

Bogor's terrain is just as dramatic as its climate, nestled here on the Western Java hills. Despite its altitude being only 290 meters, the city is surrounded by extinct volcanoes, and it was probably this combination of milder seasons, and its position above sea level, that caused the Dutch to pick this site for a botanical garden in 1817. The initial design of the grounds was produced by German-born Dutch botanist, Professor Casper Georg Carl Reinwardt. Reinwardt was interested in plants which were used by the Javanese for domestic and medicinal purposes, and collated a whole anthology of plants taken from most parts of the Archipelago, making Bogor a natural centre of agriculture and horticulture for the region.

Bogor Botanical Gardens' enlargement was steady but rapid. In 1823, when the first catalogue of plants in the Gardens were logged, only 914 species were recorded. In under 25 years, this grew to over 2,800, and now stands at 15,000. Today, the Gardens house a large collection of palms (over 400 species of the plant), as well as a wide assortment of trees, including some colossal mahogany and teak varieties.

What makes the Gardens special, however, are the curiosities dotted around that hint at their history. For example, there is a bed planted in the three colours of the Belgian flag, which followed a visit by Princess Astrid of Belgium in 1928. There is also a monument in memory of Olivia Raffles who died in 1814 (before the Gardens were actually established). Olivia was the daughter of Sir Stamford Raffles, a man whose legacy is felt throughout the whole of this South East Asian region, not least in numerous features in the Botanical Gardens themselves. Raffles was one of the primary British colonists in this area, known for establishing the foundations of what was to become modern Singapore, as well as for governing and developing Java in the early nineteenth century. Apart from the infamous *Raffles* hotel in Singapore, Sir Stamford also lent his name to a woodpecker, a butterfly fish,

and the *Rafflesia*, the world's largest flower. With a diameter of up to 105 centimetre, the Bogor Botanical Garden was home to a *Rafflesia* until the Second World War, when the Garden was neglected and many species perished.

Sir Raffles was one of the founders and first presidents of the Zoological Society in London and the London Zoo, as well as the author of the first comprehensive *History of Java*. It was his deep interest in academic study and research—this at the height of colonialism's exploration of new worlds—that left its mark on the Botanical Garden in Bogor, who Raffles was also said to have inspired, and can still be traced today. In the mid-nineteenth century, the Gardens played a part in the research and cultivation of quinine, an extract used for treating malaria and sourced from the bark on the Peruvian Cinchona tree. Botanists at the Gardens have also undertaken research on plant parasites and diseases affecting important local crops such as sugarcane.

tag: This lightweight bridge links two forest areas across a river.
opposite: Waterlilies. top left: A woodland section of the Garden.
top right: A view through ferns and palms to the fountains, pools and formal ornamental gardens.

ISRAEL
JERUSALEM BOTANICAL GARDENS

In 1954, the erection of a new campus for the Hebrew University of Jerusalem was started on the Giv'at Ram Hill in the west side of the city, and it was decided to incorporate a botanical garden on the new site, providing the University's Botany Department with a forum for research and study. American landscape architect, L Halperin, led the overall design of the Gardens, alongside a number of esteemed Israeli garden planners. Their design was guided by landscape considerations—the hillside position needed canny and expert manipulation to successfully plant the initial collection of 800 species—and revolved around botanic classifications. This was a novel curatorial approach: the conventional planning of Botanical Gardens was based on the systematic method, that is, the positioning of plants according to their scientific classifications into orders, families, genera, and species. The design of the University Botanical Garden at Giv'at Ram, on the other hand, revolved around plants being displayed in phyto-geographic sections—by the natural distribution—to illustrate different plant landscapes of the world. At this time, the botany department's distinct array of conifer trees were the highlight of the collection.

At that stage, the Hebrew University of Jerusalem was Israel's only university (it was founded in 1925 by Albert Einstein, Sigmund Freud, Martin Buber and Chaim Weizmann) and the developmental needs of the University, coupled with the special needs of the Garden, led to a decision to move the institution to a new, separate site, close to the southeastern corner of the University's Mount Scopus campus. The slow and arduous development of the new site begin in 1962 with a plantation of North American Conifers planted in the desolate, rocky soil. It would be another 15 years before the land was touched again.

During the 1960s, the Mount Scopus campus felt the brunt of the Six-Day War of 1967 and thereafter all efforts to regenerate

the new Botanical Gardens were sidelined by the need to restore the actual campus. It was not until 1975 that the development of the site was kick-started with the establishment of The Society of Friends of the Botanical Gardens. That year, a meeting was held in the office of Abraham Herman, President of the Hebrew University, with the participation of Teddy Kollek, Mayor of Jerusalem, and the Gardens' supporters. It was decided to make the Botanical Gardens a joint project of the University, the City of Jerusalem and the Jewish National Fund. This was done to divide the burden of development and maintenance among the three bodies. Also, a scientific board was appointed to guide the scientific activity and the development of the Gardens. Architect, S Aharonson, was entrusted with the planning of the Gardens. His first step was to create space for plants from South-west and Central Asia—donated by Hyman and Irene Kreitman (the latter was a famous art patron and daughter of Tesco founder, Sir John Cohen)—and to establish the Mediterranean Garden.

It was not until 1985, over 30 years after the Garden was first created, that it was opened to the public. Ever since then, the grounds have steadily grown. In 1986, F Dvorsky donated the Tropical Conservatory—still one of the Garden's most important features—to the organistion. Planting on the South African lot began in 1989, and in 1990 the development of the Dvorsky Visitors' Centre started. Today the Gardens hold more than 10,000 species—arranged in six geographical sections, representing most of the world. Special botanical attractions include The Oak Collection (70 species), The Bulb Collection (400 species), The Asian Fruit Tree Collection, The Aloe Collection, The Outdoor Bonsai Collection, a herb-medicinal garden, and the shelter garden for endangered species.

tag: A selection of the Garden's aloe plants. opposite: *Urginea maritima,* a relative of the hyacinth which has historically been used as a pesticide. above left: Iris (*Ixia viridiflora*). above right: This large variety of Aloe tree originates from South Africa.

BOTANICAL GARDEN OF THE UNIVERSITY OF PADUA

The Botanical Garden of the University of Padua was listed by UNESCO as a World Heritage Site in 1997. In their appraisal of the Garden, the organisation identified how Padua had been the first to focus on scientific study, and the first to develop links between horticultural life and scholarly research. UNESCO also identified how the Garden had deeply and historically contributed to the development of numerous Renaissance disciplines, including botany, ecology, medicine and chemistry. It was this academic reputation that the Botanical Garden of the University of Padua has always keenly maintained and it seemed fitting that, over four hundreds after its construction, it was officially awarded for employing a model borrowed frequently by botanic gardens thereafter.

The original layout of the garden was designed in 1545 by Venetian scholar and architect, Daniele Barbaro, and is still very much apparent today. Barbaro's landscape was influenced by arithmetic and geometry, and consisted of a circle within a square that was carved up by two bilateral paths. This design was largely influenced by the four cardinal directions of the compass, and is cleverly repeated in the numerous sundials that are dotted around the 22,000 square metre landscape. Architecturally, Barbaro also made the decision to implement the medieval model of the *Horti Conclusi*—or enclosed garden—with the supposed intention of keeping out nocturnal plant-thieves: government threats of imprisonment, large fines and exile apparently were not enough.

Barbaro's landscape plan has been progressively added to, ever since the Garden's inception. The latter half of the sixteenth century brought fountains and extensive irrigation mechanisms, while the eighteenth century saw the outer wall—considered sparse by contemporary decadent tastes—decked out with histrionic

gateways, balustrades and ornaments. Proving that it still retained its position as primarily an *academic* institution, as opposed to leisure or cultural centre, the nineteenth century saw many research facilities erected, included greenhouses and a botanic theatre decorated with portraits of eminent botanists in the Garden. Recent years have seen a herbarium, library and numerous laboratories added to the complex, moves that artfully affirm the Garden's reputation as a horticultural and scholarly hub.

Nowadays, this reputation is built as much on the Garden's ago as on its academic standing. Officially, The Botanical Garden of the

University of Pisa—founded in 1544—is older by just one year, although it has changed sites twice since its inception. The Garden in Padua, on the other hand, was founded in 1545 at the suggestion of the Senate of the Venetian Republic, and has remained in the same location ever since. Still growing in the Garden are some of the oldest specimens of numerous horticultural types. There is a ginkgo tree and magnolia tree, both reputedly dating back to the eighteenth century, and both of which claim to be the oldest European examples of each breed. However, it has been said that the oldest is the Chaste Tree (*Vitex agnus-castus*) that was planted not long after the Garden's inauguration in 1545. Sadly, the tree died in 1984.

In the sixteenth and seventeenth centuries, Padua's close proximity to Venice—an important centre in the world's trade market—proved massively advantageous for the Garden's abundant collection of foreign specimens. As a result, the Garden now cultivates over 6,000 types of plants, carefully arranged according to taxonomic, utilitarian, ecological-environmental and historical standards. Besides the carnivorous, aquatic, insectivorous and

poisonous plants, their collection of medicinal plants is still the most important to the horticulturists at Padua, signalling that the garden still remains true to its original didactic function.

tag: The Goethe Palm (*Chamaerops humilis*) Greenhouse. opposite: A historical illustration of the Garden. left: A fountain surrounded by statues in the Garden. above: The entrance to the Garden.

BOTANICAL GARDEN OF THE UNIVERSITY OF PISA

The Botanical Garden of the University of Pisa was responsible for formulating a model that has been borrowed by almost all subsequent botanical gardens thereafter. As it stands today, Pisa is by no means the largest or the most significant botanic garden— nor is its collection as ample as those at Kew, Berlin or Brooklyn— but its rank as the birthplace of modern-day botany is enough to position it as one of the world's most notable botanic gardens.

The Botanical Garden of the University of Pisa was founded in 1544 by Duke Cosimo I de'Medici. The Garden was one of the first examples of his regeneration and development of the Tuscan area; he later went on to construct the Uffizi Palace, as well as remodelling the Pitti Palace complex, having acquired the site through marriage in 1539. He added the renowned Boboli Gardens attached to the palace, supposedly heavily influenced by his first horticultural project in Pisa.

On establishing the idea of a garden connected to the University of Pisa, he entrusted the site to the care of Luca Ghini of Imolia, a worthy physician and recently appointed Chair of Botany at the University of Pisa. It was the first major garden in Europe to belong to a university, and was responsible for binding the close relationship between scholastic centres and gardens. Essentially, as the concept is understood nowadays, it was the world's first botanic garden. Ghini's focus was firmly placed on plants with medicinal and pharmacological properties, and he put great emphasis on the scientific value of plant life. In the sixteenth-century, just as today, the Garden was of central importance to all sectors of research and teaching in plant biology.

The Garden at Pisa enforcedly shares its authority with the Botanical Garden of the University of Padua. Although one year younger than Pisa, the Botanical Garden of the University of Padua has remained on the same site since its inception in 1545, and hence holds the title of most ancient botanic garden. Pisa, on the other hand, was shifted from its site near the church of San Vito in 1563 to the vicinity of the convent of Santa Marts. Then once again in 1591, it moved to its third and final location in via Santa Maria,

near the Cathedral Square. At this site, the Gallery of Natural and Artificial Finds and the Library were installed, and still exist in some part today. The collections and architecture of these centres form The Museo di Storia Naturala and the Library collection has since been absorbed in to that of the main University Library's.

Highlights of the Garden include the herb gardens and arboreta, and the sea-shell adorned Botany Institute, built between 1591

and 1595. There is also an extensive set of ponds, greenhouses and thematic gardens.

tag: Imperial Crown (*Fritillaria imperialis, Liliaceae*) in bloom. opposite top: Entrance to the Garden. opposite bottom: Part of the Tropical Greenhouse Collection. above: The Aloe Collection in the Succulent Greenhouse. all photographs: Dr Giuseppe Pistolesi, Curator of the Botanic Garden of Pisa.

The Botanic Garden

The history of the botanic garden can only be told through the stories of those pioneering institutions that gave the concept shape. Beginning with the first botanic gardens in Pisa and continuing through to some of the most innovative institutions working today, the curators and directors of some of the world's most important gardens here chart the history of their practice, and discuss the principles behind the botanic garden as institutions of research, conservation and recreation.

The Botanical Garden of the University of Pisa

Fabio Gabari

In 1543, the great naturalist, herbalist, and physician Luca Ghini, 1490–1556, was summoned by Grand Duke Cosimo I de'Medici, 1519–1574, from Bologna to Pisa and given a piece of land for the purpose of teaching botany. In a letter dated 4 July 1545, Ghini states that he had gathered plants "which I have planted in a garden of Pisa to be useful for the students". From this we can infer that his garden of simples was the world's first academic botanic garden. Ghini's garden was soon replaced by another in the eastern part of the city. It was entrusted to Andrea Cesalpino, 1525–1603, the most brilliant of Ghini's pupils. However, its site also proved unsuitable, and Grand Duke Ferdinand I, 1549–1609, who was Cosimo's son and successor after his brother Francesco de'Medici, 1541–1587, ordered the Garden to be moved again. During the years 1591 to 1595, 200 metres from Pisa's baptistery, cathedral, and famous leaning campanile, the third and remaining botanic garden was created. Its construction initially was entrusted to Lorenzo Mazzanga, probably a student of Cesalpino's, and then to the Flemish gardener Jodocus De Goethuysen known as Giuseppe Casabona, who had served the Medici family in Florence.

The plan of this garden, which can be seen in a copperplate engraving in the *Catalogus Plantarum Horti Pisani* published in 1723 by Michelangelo Tilli, 1655–1740, is a descendant of the Botanical Garden of the University of Pisa. It depicts eight slightly rhomboid beds, each of which is divided into smaller geometric shapes, which may have had symbolic significance of an astrological or religious nature. A circular or octagonal fountain marked the center of each bed, six of which are still in place. Tilli's *Catalogus* lists more than 4,000 plants cultivated in the Garden, 50 of which are illustrated by the artist Cosimo Mogalli. Since the end of the sixteenth century, many artists have been commissioned to illustrate the Garden's specimens and a collection of their watercolours is preserved in the central library of the University of Pisa.

Although intended to be instructive, the Botanical Garden of the University of Pisa was also a place of pleasure where members of the Medici family and their guests could discuss scientific, artistic, and literary subjects and enjoy the display of specimen plants. In addition to presenting exotic plant material, the Garden boasted a cabinet of curiosities—a gallery showcasing thousands of specimens, such as corals, minerals, whalebones, a dried crocodile, a mummy, a variety of shells, and fossil plants and animals. The facade of this forerunner of the natural history museum was decorated in the Grotesque style, and its restoration in 2005 perpetuates an important element of the Garden's early years.

Other parts of the Garden reveal the extensive revisions to the original plan undertaken in 1782 by the new director Giorgio Santi, 1746–1822. During his enthusiastic stewardship, the old beds of medicinal plants were reconfigured as a series of symmetrical rectangles and replanted with new species that were classified according to Carolus Linnaeus' system of binomial taxonomy. In addition, outside the confines of the Garden's basic geometrical layout, Santi planted an arboretum. A fine Ginkgo Tree (*Ginkgo biloba*) and a *Magnolia grandiflora*, both planted in 1787, are still standing.

Under the direction of the noted botanist Gaetano Savi, 1769–1844, the Garden gained new glasshouses and a special conservatory for aquatic plants. Savi also increased its library and herbarium collections, and in 1839 he hosted a historically

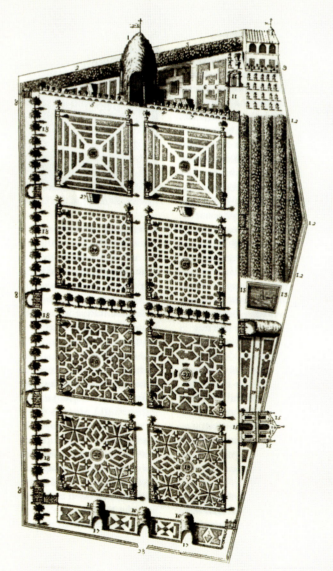

The Botanical Garden of the University of Pisa, copperplate engraving, 1723.

important first meeting of Italian scientists, himself chairing the session on botany and plant physiology. Teodoro Camel, 1830–1898, director during the second half of the nineteenth century, continued to enrich and document the Garden's collections while further revising its layout.

Giovanni Arcangeli, 1840–1921, the resolute, versatile and prestigious naturalist who succeeded Caruel, managed to acquire all the land delimited by four streets of the town, thus enlarging the Garden to its present size. He also constructed a Neoclassical botanic institute in its centre. As a skilled systematic botanist experienced in agricultural techniques, Arcangeli studied the practical applications of plant biology. In addition, he was a highly respected taxonomist. His 1892 *Compendia della Flora Italiana* outlines the modern concept of subspecies.

The three hectare garden is divided into two principal parts: the southern half contains the school and the northern half the arboretum. In addition, 880 square metres are devoted to glasshouses and service areas. In order to facilitate maintenance and to improve botanical pedagogy, the school section recently has

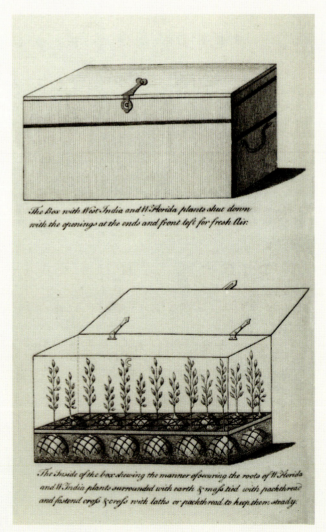

Early examples of plant containers used in collecting expeditions during the eighteenth century.

been subdivided into smaller beds, each containing a single herbaceous species, mainly Mediterranean ones. Students and teachers from the University of Pisa in the courses of biology, natural sciences, environmental sciences, agriculture, veterinary medicine, and pharmacological studies regularly visit the Garden. 12 staff members—the director, the curator, and ten gardeners—ensure that it contains material for their courses and laboratory experiments.

In addition to fulfilling its traditional didactic and scientific roles, the Botanical Garden of the University of Pisa now focuses on the conservation of plants threatened with extinction. A humidity-controlled seed bank with a temperature of 20 degrees below zero has been provided to store seeds of critically endangered or vulnerable species gathered mainly from the National Park of the Tuscan Archipelago, the Regional Park of the Apuan Alps, the San Rossore Estate, and their surroundings. Scientists from the University's biology department study their physiology in order to develop appropriate strategies for ex situ conservation. Further, the Garden hosts the presidency of the Rete Italiana Banca Ex Situ a national network of approximately 20 research units dealing with the conservation of Italian flora. In addition, it is a member of the European Native Seed Conservation Network. The twin aims of both organisations are to promote the quality, coordination, and integration of European native plant conservation practice, policy, and research and to assist the European Community in meeting its obligations to the Rio de Janeiro Convention on biological diversity. Thus, the world's oldest botanical garden, while rich in historical material—much of which is displayed in its fine museum containing portraits of famous early botanists, wax models of more than a hundred fungi, and the 'studiolo' or multi- drawer writing desk, of Grand Duke Ferdinand I (the seed bank of its day)—is also a modem institution whose distinguished past is linked to the challenges of the future.

Hortus Botanicus Leiden

Gerda van Uffelen

Portrait of Carlous Clusius, 1526–1609, Flemish doctor and perhaps the most influential of sixteenth century horticulturists.

The Sixteenth Century

Founded in 1590 by the curators of Leiden University on a 30 metre square plot obtained from the municipality at the back of their academy, the Hortus Botanicus Leiden is the oldest botanic garden in The Netherlands. Like other early botanic gardens, its original purpose was to instruct medical students on the healing properties of various plants. In 1592, Carolus Clusius (Charles de l'Escluse), 1526–1609, a major figure in Renaissance botanical science who recently had laid out a garden in Vienna for the Emperor of Austria, agreed to come to Leiden to become its first prefect, or scientific director. He brought with him a large tulip collection that eventually was planted in Leiden, thus forming the basis of the tulip trade in The Netherlands.

Clusius, then age 66, had traveled widely all over Europe and had published extensively. He continued to maintain a vast network of scientific correspondents. Because of his advanced years and his having been seriously injured by a fall, the University appointed Dirck Outgaertszoon Cluyt, 1546–1598, a pharmacist from Delft, as his assistant. Cluyt, or Clutius according to his Latin appellation, was called *hortulanus*, the keeper of the Garden. Working as a team, Clusius and Cluyt developed a plan with carefully numbered beds accompanied by a list of the plants they intended them to contain. In 1594 they presented this plan to the overseers of the University, who were surprised to find that what Clusius and Cluyt envisioned was a botanic garden laid out for the study and enjoyment of plants rather than a simple herb garden focused solely on *materia medica*. Nevertheless, the Garden was constructed according to their intentions.

The Seventeenth Century

In the summer of 1600 the Ambulacrum, the Hortus Botanicus' first permanent building for the protection of delicate plants in winter, was constructed. Here, on the south side of the Garden, both plants and students could find a place sheltered from inclement weather. The Ambulacrum might be called the oldest museum in The Netherlands because of its collection of curious objects, including representations of a dragon and a mermaid described in three contemporary inventories.

The oldest plant still surviving from that early period is the Golden Chain Tree (*Laburnum anagyroides*) next to the main entrance. It was planted in 1601, at which time Clusius was preparing two of his major works for publication: *Rariorum Plantarum Historia*, 1601, and *Exoticorum Libri Decem*, 1605. At the same time he wrote letters to the board of the Dutch East India Company encouraging its members to ensure that physicians and other travellers to faraway places collected exotic plant material for study in the Hortus, an activity continued by his successor, Pieter Pauw, a professor of botany at Leiden University. Between 1669 and 1676, Antoni Gaymans, a pharmacist in Leiden, accumulated a still extant large herbarium containing more than 1,450 foreign plants, many of which became part of the collection of the Hortus. In this way the collection grew spectacularly over the decades from 1,000 species in 1594 to 3,000 by 1685 when Paul Hermann, also professor of botany at Leiden University, was prefect of the Hortus.

The Eighteenth Century

From 1709 to 1730, Herman Boerhaave, a physician of worldwide standing, was the director of the Garden. A catalogue published two years after his death in 1738 lists approximately 7,000 species. In the eighteenth century several exotic trees were planted that still survive. These include a Tulip Tree (*Liriodcndron tulipifera*), from North America planted sometime between 1710 and 1720, a Date Plum (*Diospyros lotus*) from Asia planted around 1740, and a Ginkgo (*Ginkgo biloba*), a tree Europeans had previously thought extinct but which had been rediscovered in China a few years earlier, planted in 1785. The eminent Swedish botanist Carolus

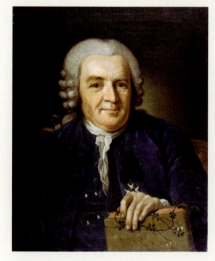

Carolus Linnaeus, 1770–1707, pioneering Swedish botanist who invented modern botanical taxonomy.

Linnaeus met Boerhaave and visited the Hortus during his stay in The Netherlands from 1735 to 1738, during which time he reputedly planted a specimen of the Alpine Honeysuckle (*Loniceri alpigena*).

Following some small additions in the seventeenth century, the Garden was substantially enlarged in 1730, covering land on both sides of the Binnenvestgracht, the canal that continues to run through it. Its size at this point was approximately 1,600 square metres. A large brick orangery, designed by the French Huguenot architect Daniel Marot, 1661–1752, was erected in 1744. This building housed both a large number of tub plants in winter and a collection of classical sculptures left to the University by Gerard van Papenbroeck, founder of the University of Amsterdam.

The Nineteenth Century

During the heyday of the exploration of the Dutch East Indies, the search for and study of useful and valuable plants led to the establishment of the Rijksherbarium, or National Herbarium, in 1829 by royal decree of King Willam I. Today its Leiden branch houses about 4,000 specimens. In 1816 the Garden quadrupled in size when the city bastion beside the Singel Canal was extended. The new part of the Hortus was laid out in the then popular *Jardin Anglais* style, and several of the remaining trees from that period are among its most venerable specimens. These include an enormous Horse Chestnut (*Aesculus hippocastanum*), a Caucasian Wingnut (*Pterocarya fraxinifolia*), and a Fern-Leaved Beech (*Fagus sylvatica* 'Asplenifolia'). Many of the Garden's nineteenth century trees were imported from Japan by Franz Philipp von Siebold, 1796–1866, a German physician in the service of the Dutch East India Company. He collected and described a large number of animals and plants as well as an enormous number of Japanese objects, which are now exhibited in the Ethnology Museum in Leiden and in the Siebold House, which is located near the Hortus. About 15 plants in the Hortus were personally imported by von Siebold, who lived in Leiden for several years and even owned a nursery there. He introduced many well-known garden plants such as hydrangea and wisteria into Holland. In 1990 a newly-designed Japanese garden commemorating von Siebold was laid out around a *Zeikova serrata* tree that he had planted.

The recreated Clusius Garden in the Hortus Botanicus Leiden. Image courtesy of the Hortus Botanicus Leiden.

In 1857 the Hortus had to give up a parcel of land in order for the university to build an observatory. Today its telescopes are no longer employed for scientific purposes, and it may prove possible for the Garden to repossess the observatory grounds and gain more space in which to increase its outdoor collections.

The Twentieth Century

In the last century many changes were made to the Hortus that are still visible today. Between 1930 and 1940 the prefect, professor Lourens Baas Becking, and the *hortulanus*, Hesso Veendorp, collaborated on the completion of some large projects begun earlier. They laid out a new rose garden, built a new set of tropical glasshouses to replace a number of glasshouses scattered throughout the Garden, and undertook a replication of the 1594 garden laid out by Clusius. Since 1999 the united Hortus and the Rijksherbarium have functioned as a separate institute of Leiden University. Its combined collections have grown, and 77 per cent of the now more than 12,000 specimens are used for research and teaching.

The glasshouses contain the larger part of the collections—mainly tropical plants from Southeast Asia, (specially orchids, ferns, pilcher plants, and cycads). These form part of the national plant collection, established in 1988 in which 17 Dutch botanic gardens participate. Contemporary taxonomic research includes DNA analysis, for which purpose a laboratory is situated next to the Hortus.

The Twenty-first Century

Since the turn of the twenty-first century, the Hortus Botanicus Leiden has gained the Winter Garden, a temperate greenhouse for subtropical plants. In 1993 Gerda van Uffelen, a botanist who has studied fern spores and published on ferns, was appointed collection manager with responsibility for horticultural administration and, beginning in 2005, the laying out of a new systematic garden according to the latest results of plant DNA.

Research concerning the early years of the Garden continues and recently was benefited by the discovery in Krakow of a set of several hundred botanical watercolours dating from the second half of the sixteenth century. Many of the plants found in these *libri picturati* were present in Clusius' original layout of the Hortus, and there is reason to believe that the watercolours were made in Flanders under his supervision. In addition, there are close links between the Hortus, the National Herbarium of The Netherlands, and numerous museums in Leiden, such as the Naturalis, the Museum of Natural History. Today the Garden is a haven of rest in the middle of a university town and attracts students and visitors from all over the world. It is a registered museum, where many Leiden citizens have queued at night to witness the large nocturnal bloom of the Giant Waterlily (*Nymphaea amazonia*) or to have their babies photographed on one of its enormous floating leaves. All schoolchildren in and around Leiden are offered a visit to the Hortus during their school career. Thus, this remarkable 416 year old botanic garden continues to play an important role in the life of both the University and the city of Leiden.

Chelsea Physic Garden

Rosie Atkins

In 1673 the Society of Apothecaries of London founded a Physic Garden at Chelsea so that its apprentices could learn to grow medicinal plants and study their uses. Similar teaching gardens were created in Padua and Florence, and the universities in Bologna, Leiden, Montpellier, Edinburgh, and Oxford soon founded others. With the exception of the Chelsea Physic Garden, these medicinal teaching gardens grew into botanic gardens as we know them today. Never having been attached to a university or teaching hospital, Chelsea maintains continuity with its origins.

When the Society of Apothecaries chose to rent four acres beside the River Thames, the area known as Chelsea consisted of green fields, market gardens, and orchards. London was still recovering from the Great Fire of 1666 and several years of plague. Travel was infinitely safer and quicker by boat than by road, and King Henry VIII, his chancellor Sir Thomas More, and Sir John Danvers had all built fine country houses in Chelsea. The location, which was a convenient distance from the crowded city and had the added attractions of good, free-draining soil and a southerly aspect, also met the Society's need for a place near the river to house the gaily painted barge used for royal pageants and for their celebrated 'herborising' expeditions to collect plants.

The Chelsea Physic Garden as seen from across the Thames in a mid-eighteenth century painting after Canaletto, 1697–1768. The boathouses, centre, were intended for the boats used in 'herborising' expeditions to collect plants.

For the first ten years, the Society had difficulty finding a good gardener to grow simples, the herbs that the apothecaries who were its members would have taken downriver to their guildhall at Blackfriars. However, by 1683 John Watts, the apothecary whom they appointed to oversee the Garden, was able to establish valuable links with Paul Hermann, a professor of botany at Leiden University, and the two men were exchanging plants and seeds, the most famous being four seedlings of the Cedar of Lebanon (*Cedrus libani*), which had never before been cultivated in Britain. Offspring of the Chelsea Physic Garden cedars can still be found in botanic gardens and old estates, and the Garden continues to exchange seeds with other botanic gardens around the world and to publish a yearly *Index Seminum*. As early as 1685, the celebrated diarist John Evelyn describes a heated glass house, thought to be the first in Europe, along with one of the plants it sheltered, a Cinchona Tree (*Cinchona ledgeriana*), the source of quinine, a drug promoted by the physician Hans Sloane, 1660–1753, an important figure in the Garden's history. Sloane, who studied medicine at Montpellier in France and who was appointed president of both the Royal Society and the Royal College of Physicians, was knighted in 1716. By 1712 he had acquired enough money to buy the manor of Chelsea, and in so doing, he also took over the freehold of the Chelsea Physic Garden. Sloane was sympathetic to the Society's constant struggle to pay the rent on the property, and in 1722 he granted them a lease of five pounds a year in perpetuity on condition the Garden "be for ever kept up and maintained as a physick garden".

Sir Hans Sloane, 1660–1753, without whom the continued success of the Chelsea Physic Garden would not have been possible.

This deed of covenant secured the Garden's future and established its place in horticultural history. One condition of the lease required that each year the garden must deliver 50 pressed and mounted plant specimens to the Royal Society until 2,000 had been received. By 1795 the Garden had provided more than 3,700 herbarium specimens, which are now housed at the Natural History Museum in London.

When Sloane died at the age of 92, his collection of curiosities and his vast library became the nucleus of the British Museum. His plant specimen collection was later moved to the Natural History Museum. Botanists from this institution continue to use the Chelsea Physic Garden and its team of expert gardeners to help them with their research. Sloane's name lives on in such local landmarks as Hans Crescent and Sloane Square as well as in the fixed rent of five pounds that the Chelsea Physic Garden still pays to Sloane's heirs every year.

On Sloane's recommendation, Phillip Miller, 1691–1771, a Scottish botanist, was appointed head gardener in 1721. Miller, who made the Garden world famous during his 50 year tenure, trained William Atton, 1731–1793, another Scottish botanist and the first director at the Royal Botanic Gardens at Kew, as well as his successor at Chelsea, William Forsyth, 1737–1804. Miller's correspondence with the leading botanists of his day generated an exchange of plants and seeds, many cultivated for the first time in Britain. Miller also introduced eight editions of his famous *Dictionary of Gardening*, which became the standard reference work for several generations of gardeners in Britain and America.

Carolus Linnaeus, the great Swedish botanist who is considered the father of binomial Latin plant taxonomy, made several visits to the Garden in the 1730s, and many species first described by Miller retain the names Linnaeus ascribed to them.

Sloane was active in fostering economic botany-research on cash-producing crops—an endeavour promoted by Miller, who arranged for various crops including cotton to be sent out from Chelsea to the new colony of Georgia in America. Miller also introduced the cultivation of Madder (*Rubia tinctorum*), the roots of which are used to produce red dye, as an agricultural crop in Britain. In 1732 Sloane laid the foundation stone for an orangery where Miller lived for a short time with his family. Sadly, this elegant building was pulled down in the mid-nineteenth century when the Chelsea Physic Garden's fortunes went into decline.

In 1899 the Society of Apothecaries finally gave up the management of the Garden, and it was taken over by the City Parochial Foundation. Until 1983 it remained closed to the general public, although university and college students continued to use it for scientific research. When the City Parochial Foundation then determined that it could no longer maintain the Garden, a new independent charity was established to manage and operate it. At this time it was also decided to open the Garden to the public for the first time in its 300 year history.

Today the Chelsea Physic Garden occupies 3.8 acres of prime London real estate bounded by Royal Hospital Road to the north, Swan Walk to the east, and the Embankment to the south. As the Embankment is a busy thoroughfare, the Garden is now cut off from the Thames. Otherwise, it has changed very little since the mid-eighteenth century. The main building's offices, lecture rooms, and the curator's house remain—as well as most of the glasshouses at the northern end of the Garden. Gravel paths divide it into quadrants, and grass paths run between beds that are planted in a manner that demonstrates the botanical relationships of various plants. Beds in the northeast quadrant display plants used in the pharmaceutical industry as well as plants such as the Opium Poppy that have been used over the centuries in herbal medicines. In addition, there are culinary plants, ones for the perfumery and cosmetic industries, and others that are used in the manufacture of fabrics and dyes. In 1993 the Garden laid out *A Garden of World Medicine*, a living exhibit displaying medicinal plants used by the world's indigenous peoples.

A replica of the 1733 statue of Sir Hans Sloane by Michael Rysbrack, 1694–1770, stands in the centre of the Garden. The original, which was being damaged by air pollution, is now in the British Museum. Next to the statue is an exhibition created in 2003 to commemorate the 250th anniversary of Sloane's death. Nearby is the oldest rock garden in Europe. Here the rocks include pieces of the Tower of London and basalt used as ballast on Sir Joseph Banks' ship on a voyage to Iceland in 1772. In the northeast corner an education building, opened in 1997, is used to teach children about the vital role plants play in our lives. Nearby is the Historical Walk, which charts the Garden's history with plants introduced into cultivation over the centuries by the Garden's curators and by other notable botanists including Banks,

William Hudson, 1730–1793, William Curtis, 1746–1799, and Robert Fortune, 1812–1880. By the Embankment a wider area of flowering shrubs and rare peonies offer places to sit and enjoy the Garden's tranquil atmosphere.

Wildlife flourishes in the Garden. The frogs, toads, and newts inhabiting the Fortune's Tank Pond help control the growth of the slug population. This year video cameras installed in the Tank Pond and a bird box have broadcast wildlife activity live on television. *The Ethnomedica Project*—a joint initiative with the Royal Botanic Gardens at Kew, the Eden Project, the Royal Botanic Garden in Edinburgh, and the Natural History Museum supports the collection of data on herbal remedies that have been used over the centuries in Britain. In June 2006 Princess Alexandra opened the Back to the Garden recycling project, which reveals the mysteries of making compost and the recycling of green waste. Like all botanic gardens in the twenty-first century, conservation, scientific research, and education play vital roles in the Chelsea Physic Garden's activities. Indeed, the Garden can be said to be London's oldest outdoor classroom.

British Colonial Botanical Gardens in the West Indies

Nina Antonetti

The rapacious exploitation of the New World colonies for plantation development and exportation of cash crops in the eighteenth century led scientists to recognise nature as both bountiful and fragile. British colonial botanic gardens played an important role in achieving this understanding

The history of British colonial botanic gardens can be traced back to the beginnings of botanical science in England. John Parkinson, 1567–1650, apothecary to King James I of England and a founding member in 1619 of the Society of Apothecaries, and John Evelyn, 1620–1706, English diarist, author, gardener, and early environmentalist, were among those who fostered the development of modern botanical science. Parkinson, the author of *Paradisi in Sole Paradisus Terrestris*, 1629, and *Theatrum Botanicum*, 1640, was able to graduate from the role of herbalist to that of botanist and to examine and write comprehensively about plants imported from exotic lands. Evelyn, one of the founders of the Royal Society in 1660, is the author of two important treatises: *Fumifugium* or *The Inconvenience of the Aer and Smoak of London Dissipated*, 1661, and *Sylva, or Discourse on Forest Trees*, 1664.

The further progress of botanical science during the Enlightenment was characterised by a zealous search for hitherto unknown plants. Several plant-hunting expeditions were sponsored by the Royal Society under the direction of the noted naturalist, botanist, and trusted scientific adviser to King George III, Joseph Banks, 1743–1820, and by the Royal Botanic Gardens at Kew under the direction of the Scottish botanist William Aiton, 1751–1793. Banks himself went on voyages to Australia, the Faroes, and Orkney Island in Scotland, discovering nearly 80 species. In 1789 Aiton published *Hortus Kewensis*, a catalogue of all the plants in cultivation in the Royal Botanic Gardens. The explorer, navigator, and cartographer Captain James Cook, 1728–1779, also played an important role in this era of discovery, conferring the name Botany Bay on the harbour where Banks and the Swedish botanist Daniel Solander, 1733–1782, who were attached to his voyage to Australia, enthusiastically collected numerous plant species. Soon colonial botanical gardens began to be created for the collection and propagation of plants intended for shipment to Kew. These played a central and sustaining role in colonial botanical expansion during the directorships of William Jackson Hooker, 1785–1865, and Joseph Dalton Hooker, 1817–1911, father and son. With ties to Britain and a strong link to India, Kew-sponsored colonial botanic gardens provided a network of scientific enquiry and economic trade.

As British colonisation proceeded, information began to flow back home in the form of diaries, correspondence, ledgers, and sketches. These provide historians today with precise and valuable evidence of the material, economic, political, cultural, and scientific world during England's age of imperial expansion. Botanical illustrators were among those sailing back and forth from motherland to colony. The naturalist and artist Marianne North, 1830–1890, for whom the Marianne North Gallery at Kew is named, was one of a handful of women central to the history of the colonial botanic garden. During her travels to the West Indies, as well as to

View of the Botanic Garden of St Vincent from the superintendent's house. Lithograph of a drawing by the Reverend Lansdown Guilding in his *Account of Botanic Garden in the Island of St Vincent*, 1825.

Brazil, Java, India, Africa, and Australia, she documented more than 900 species in more than 800 botanical paintings.

Paradoxically, as the ancient ideal of paradise continued to expand in European literature, drama, and art, and while the realms of research science and horticulture were being vastly enriched by the importation of exotic species, native plant communities in the British West Indies were being degraded. This was especially true on the islands of Barbados and Jamaica, where the clear-culling for sugar plantations by slave labour caused rapid deforestation and resulting soil depletion. As a consequence, eighteenth and nineteenth century botanists there found themselves engaged in issues of soil erosion and depletion of native species coupled with famine and environmentally induced illness. Thus, while the first colonial botanic garden managers were mainly interested in shipping specimens back home, their successors began to provide reports of deforestation, soil erosion, and land mismanagement. Evelyn's plea in *Sylva* for the reforestation of England, where extensive shipbuilding had depleted its timber resources, fell on deaf ears at home; however, his concerns became obvious in colonial lands where the effects of aggressive tree cutting on soil and climate were more obvious. As a result, new conservation restrictions were made to benefit such island settlements as Tobago, which consequently escaped the environmental destruction inflicted on Barbados and Jamaica.

Founded in 1754, the Royal Society of Arts subsequently played a key role in launching measures to protect the natural resources of British territories. In 1765 the Society assisted in the creation of the first colonial botanic garden in the western hemisphere on St Vincent in the Caribbean. The medically trained Scottish botanist James Anderson was its first curator. Among the tropical species still protected there is the Breadfruit Tree (*Artocarpus altilis*), which was brought to the island from Tahiti in 1793 as a potential food crop for slaves by Captain William Bligh, 1754–1817, whose name is associated with the famous mutiny aboard the *HMS Bounty*. Anderson launched a bifurcated campaign, balancing fact-gathering missions with interest in native cultures. Despite the depredations inflicted by the Arawak and Indian Carib tribes and then by the British and French along with the volcanic eruptions of Soufriere in 1812 and 1902, the St Vincent Botanical Garden has survived and retained its reputation for forest protection and the conservation of rare species.

The colonial botanic gardens of the West Indies always have been threatened with extinction. More modest in scale and reputation than the great sponsoring gardens in Britain, they are still plagued by natural and political storms as well as by chronic underfunding. As an example of their tenuous state, visitors arriving at the grounds of the Dominica Botanic Gardens at Roseau in the eastern Caribbean are confronted with a startling scene: a school bus crushed by a fallen Baobab tree (*Adansonia digitata*) during Hurricane David in 1979 is now part of an exhibit demonstrating how prostrate trees put forth new growth.

The ecological destruction and the loss of biodiversity in colonies during the period of imperial rule is now a global phenomenon. The environmental concerns

The Avocado (*Persea gratissima*) was one of the many specimens that found its way into exploration ships during the various expeditions by European colonial powers.

1.

2.

3.

5.

4.

originally raised on such islands as St Vincent, Tobago, Trinidad, Dominica, Barbados, and Jamaica, have become universal. In spite of their lack of financial resources, the British colonial botanic gardens in the West Indies are participating in land restoration projects that are germane to the regeneration of plant species throughout the world. Thus, the recovery of parts of these ecologically damaged islands offers lessons and hope for the rebuilding of native plant communities elsewhere.

The Cambridge University Botanic Garden

John Parker

In 1831, on the wheat fields to its south, the University of Cambridge established a new botanic garden to replace its small physic garden, founded in 1762, which lay in a smoky location overshadowed by buildings in the heart of the city. The new garden's design, planting, and care became the lifework of John Stevens Henslow, 1796–1861, an accomplished mathematician, zoologist and artist, as well as an ordained priest. A professor of mineralogy from 1822 until his resignation in 1827, he also served as the University's professor of botany from 1825 until his death.

Henslow considered trees to be the most important plants in the world and was particularly excited by the recent discoveries of such conifers as the Douglas Fir (*Pseudotsuga menziesii*) in North America. As a result, the visionary young professor planned the new botanic garden as an arboretum, thereby deflecting botanical study at Cambridge towards trees rather than the medicinal properties of plants, its original focus. Within the Garden, according to Henslow's contemporary park-like design, trees, shrubs, and herbs were to be planted in naturalistic groupings, and modern glasshouses were to extend the range of climates and, hence, plants available for study. Then as now, the Garden was meant to be a working collection in which all plants would be grown and maintained primarily for scientific research and teaching purposes.

John Stevens Henslow, 1796–1861, well-respected botanist, explorer and founder of the Cambridge University Botanic Garden.

Henslow was a passionate advocate of universal education. He was closely associated with the development of mechanics institutes, night schools for the education of working men, and the organisation of village schools for the children of illiterate farm workers. Thus, his garden for botanical research was intended to be a public institution accessible to both town and gown, a place created according to the highest horticultural standards while also providing a beautiful, tranquil haven available to everyone for recreation and education.

A specimen storage box developed to protect plants on long journeys back from early botanical exploration expeditions.

Henslow quarreled vehemently and passionately with the ruling Senate of Cambridge University for funds to develop and support his new botanic garden, and this argument eloquently expressed his view of the significant value of the botanic garden to the whole of society. However, an appreciation of the scientific philosophy that underpinned his design and planting of the Cambridge Botanical Garden has been lacking until recently. Unfortunately, as far as we know, Henslow never wrote down his theories on why he grouped trees according to certain relationships to one another, nor do we have any documentation of his short-lived curator, Andrew Murray, who planned and planted much of the Garden. Nevertheless, a partial understanding of the intellectual framework guiding its development recently has emerged through close scrutiny of Henslow's scientific career and observation of the way in which the groups of trees remaining from his original plantings are interrelated.

Henslow's botanical research during the 1820s was directed toward understanding the nature of species based on the study of plant variation through the examination of wild populations and by experimental manipulation of plants in cultivation. Contemporary analysis of his herbarium specimens, research papers and letters has revealed three elements that were fundamental to the development of his understanding of species: the nature and extent of continuous variation that characterise species in nature; the phenomenon of 'monstrosity', sudden changes of flower or leaf form due to mutation or developmental abnormality; and the properties of hybrids, which he believed would reveal the laws that govern nature. Although many of the Garden's original plant specimens have been lost over the years, sufficient numbers remain to enable us to deduce some of the ideas behind Henslow's initial plantings.

Remarkably, the three themes around which Henslow focused this collection—variation, monstrosity, and hybridisation—are all represented by groups of trees still living in the Garden. So far, nine different assemblages have been identified within the surviving trees that illustrate his research, allowing us to interpret some of the interesting visual dialogues he set up among trees.

The Garden's central axis is an east-west avenue flanked by conifers. Among these is a group of four subspecies of the widespread European species of Black Pine (*Pinus nigra*). Extreme variants of this species are planted opposite each other, presenting an oddly unbalanced juxtaposition for such a majestic vista since, unlike traditional axial allees, the trees that line it are not uniform in appearance. Thus, *P nigra 'nigra'* from central Europe, with an almost unbranched growth habit and densely crowded, very short needles clustered at the apex and on the ends of thin downward-directed branches, is placed opposite a huge dominating specimen of *P nigra 'salzmannii'* from the Pyrenees, which has massive upward-directed, trunk-like branches, an open, spreading crown, and long, flexuous needles. Here Henslow's fundamental scientific query is starkly revealed: do these trees belong to one species or two? By placing other variants nearby, the commonalities of the two dissimilar pines become clear. Thus, we can ascertain that Henslow was attempting to explore visually a principle that modern botanical science confirms: we are viewing variation within a single species. Similar visual arguments are represented by plantings of the Cedrus species *Libani*, *Atlantica*, and *Deodara*, regarded by Henslow as belonging to a single species and thereby illustrating his theory of continuous variation.

Elsewhere, on the eastern side of the Garden's perimeter belt of deciduous trees, we can see Henslow's investigation of monstrosity within a group of three European beeches (*Fagus sylvatica*), which were arranged according to the 'natural taxonomic order' of the Swiss botanist Augustin Pyrames de Candolle, 1779–1841, director of the botanic garden at the University of Montpellier. One beech is a standard tree typical of British woodlands, the second is a weeping form grown from a graft on a standard rootstock, and the third is a superb specimen of the Cut-leaved Beech (*F sylvatica var. asplenifolia*) with fine-filigreed leaves instead of normal ones of simple ovate outline.

Henslow's interest in hybridisation is evident nearby in a collection of three trees of the genus *Platanus*: *P orientalis* and two different interspecific hybrids with the American Sycamore, *P occidentalis*, referred to as *P x acerifolia* (the London Plane), and *P x acerifolia 'cantabrigiensis'* (the Cambridge Plane). In addition, parents and their hybrids are still represented by some of the Garden's remaining oaks, *Quercus robur*, *Q petraea*, *Q cerris* and *Q suber*. By 1829 Henslow had a coherent view of the nature of species, which he transmitted to his students through lectures and field classes. His most assiduous student was Charles Darwin, 1809–1882, known around Cambridge as 'the man who walks with Henslow' due to his constant proximity to the professor. It was Henslow's methods of investigation and intellectual position on speciation that Darwin took with him on his epic five year voyage on the *HMS Beagle*, during which he collected plant specimens as well as those of rocks, fossils, and animals. Today one can see all Darwin's pressed plants from this expedition on sheets bearing his name in the University's herbarium collection.

Henslow's successors did not value, or did not comprehend, his innovative botanical approach, so the emphasis in the Botanic Garden shifted from plantings based on plant variation to studies in ecology and genetics as these sciences developed in the late nineteenth century. The twentieth century expansion of Cambridge swept over the countryside, and the Cambridge University Botanic Garden is now surrounded by the city. However, its boundaries have remained inviolate, and the 40 acre garden is now a serene urban oasis where many of Henslow's trees live on as testimony to the scientific brilliance of Darwin's mentor.

United States Botanic Garden

Holly Shimuzu

Visitors to the National Mall often are surprised to see a large conservatory and surrounding gardens situated so near the United States Capitol. It was President George Washington—himself the designer of the Garden at Mount Vernon—who initially envisioned a botanic garden at the seat of government. Washington wrote a letter in 1796 to the Commissioners of the District of Columbia asking that a botanical garden be incorporated into the plan for Washington, DC. Recognising the value of plants to the well-being of the young nation, he suggested that the proposed botanic garden be placed prominently in the new city and pointed out several possible sites, including the square next to the President's House.

In 1816 a group of respected citizens founded the Columbian Institute for the Promotion of Arts and Sciences. One of the institute's goals was to create a centre for scientific pursuits. The first objective of its constitution was "to collect, cultivate, and distribute the various vegetable productions of this and other countries, whether medicinal, esculent, or for the promotion of arts and manufactures". The Columbian Institute received a Congressional charter on 20 April 1818 and after considerable lobbying by its members, on 8 May 1820, Congress approved a bill providing for the use of five acres on the Mall for a national botanical garden. The bill was signed by President James Monroe and Speaker of the House Henry Clay, and the President, who accepted the title of Patron of the Columbian Institute, agreed to let the institute place the botanical garden on property adjacent to the west side of the Capitol. Other early members of the institute included presidents John Quincy Adams and Andrew Jackson, who served ex officio during their terms of office. Honorary members included former presidents John Adams, Thomas Jefferson, and James Madison.

While the collection of plants and seeds continued, work on the site began by clearing and draining the soggy land followed by tree planting. In 1824 one of the institute's members, William Elliot, wrote *A List of Plants in the Botanic Garden of the Columbian Institute*, which contains more than 100 species. In 1826 Congress appointed a committee to meet with the heads of government departments to help solicit "all subjects of natural history that may be deemed interesting" from foreign representatives. However, Congressional support was limited and maintenance of the Garden was sporadic, often done by volunteers or by an occupant of the house located on the grounds. Occasionally, the gardener from the Capitol grounds would help out after hours. The Columbian Institute for the Promotion of Arts and Sciences disbanded in 1837 due to lack of professional leadership and lack of financial support. It was reconstituted in 1941 and merged with the Historical Society of Washington.

The efforts to create the United States Botanic Garden gained momentum in 1842 when the United States Exploring Expedition with six naval vessels captained by Lieutenant Charles Wilkes, 1798–1877, returned after four years of exploring the lands along the South American, Australian, and Asian coasts; 280 islands of the South Pacific; 100 miles of the Oregon coastline and a 100 mile stretch of the Columbia River. Included on the expedition were naturalist Charles Pickering, horticulturist William Brackenridge, botanist William Rich, and geologist James Dana as well as taxidermists, artists, and a philologist. After encircling the globe and logging more than 8,700 miles, Captain Wilkes returned with 4,000 ethnographic objects and 50,000 specimens of 10,000 species of pressed plants. A place was needed to care for these botanical treasures collected by Pickering and Brackenridge.

Initially, the expedition's plant collections were housed at the United States Patent Office where a glasshouse was added to the back of the building to accommodate the study and propagation of plant specimens. However, the presence there of so much exotic flora rekindled congressional interest in having a national botanical garden, and in 1850, when the Patent Office building was enlarged, Congress appropriated $5,000 to build a new glasshouse on the site of the former Columbian Institute's previous garden. This small Gothic structure filled with rare plants quickly became a public attraction, and by the end of that year, the old garden grounds had been reestablished on ten acres of the Mall adjacent to the Capitol. Officially named the United States Botanic Garden in 1856, the Garden was placed under the jurisdiction of the Joint Committee on the Library of Congress and was given regular funding to support its growth.

Brakenridge, the horticulturist who had collected many of the plants to be installed in the reestablished garden, was put in charge. In 1853 he hired a young Scotsman, William R Smith, to begin work as gardener. Having been trained at the Royal Botanic Gardens at Kew, Smith brought experience and determination to his new position and initially was charged with preparing a comprehensive catalogue of the Garden's plants. While the majority of the plants in the Garden's collection were from the United States Exploring Expedition, Brackenridge obtained a wide variety through exchanges with other botanical gardens.

When Commodore Matthew Perry, 1794–1858, having opened Japan to Western trade two years earlier, returned from his second voyage in 1855, new species of Asian flora were added to the United States Botanic Garden. Larger glasshouses were built to display the expanding collections and to study and propagate new plants. Smith was appointed first superintendent of the United States Botanic Garden in 1863, a post he held until his death in 1912. During his tenure, the Garden experienced tremendous growth and increasing national prominence.

Built in 1867 the conservatory's rotunda contained more than 300 majestic palms in addition to plants from Asia, New Zealand, Madagascar, Panama, and South America. The wings of the conservatory housed plants from the East and West Indies, the South Seas, and China. In a nearby conservatory a lecture hall holding up to 100 people doubled as a botanical classroom.

Although well established and surrounded by lush gardens and large trees, the site of the United States Botanic Garden at the east end of the Mall became problematic at the beginning of the twentieth century when the Committee on the District of Columbia headed by Senator James McMillan, 1838–1902, sought to restore Pierre L'Enfant's, 1751–1825, 1791 plan for the nation's capital according to the tenets of the City Beautiful movement. In 1902 the McMillan Commission—a distinguished group of professionals including architects Daniel Burnham and Charles F McKim, sculptor Augustus Saint-Gaudens, and landscape architect Frederick Law Olmsted Jr—presented its report. Among its many recommendations was that the United States Botanic Garden be relocated in order to reestablish the Mall's original axis between the Capitol and the grounds adjacent to the Washington Monument, with a further extension to a grand terminus at the proposed site of the Lincoln Memorial.

Public outcry was enormous. Washingtonians, including members of Congress, were openly opposed to the move because it meant uprooting many magnificent

The United States Botanic Garden pictured in 1874, shortly after its inauguration as an official botanical institution.

trees. When the relocation from the centre to the edge of the Mall bordered by Maryland Avenue and First Street finally occurred 20 years later, more than 200 trees were destroyed and the glasshouses dismantled.

In November 1931 the cornerstone was laid for the present United States Botanic Garden's new conservatory. The following year the fountain created by French sculptor Frederic-Auguste Bartholdi, 1834–1904, for the 1870 Philadelphia Centennial Exhibition (at the same time he was working on New York City's Statue of Liberty) was brought out of storage and placed in the Frederic-Auguste Bartholdi Park, which this part of the relocated United States Botanic Garden has been called since 1985. Although now significantly smaller in size, the Garden was able to successfully continue its operations throughout the twentieth century, and in 1990 the conservatory received a major reconstruction. The newest addition to the United States Botanic Garden is the National Garden made possible by private donations to the National Fund for the United States Botanic Garden. This three acre garden is on the land adjacent to the west gallery of the conservatory. It consists of a Regional Garden, Rose Garden, and the First Ladies Water Garden.

The Elgin Botanic Garden existed on the site of the present-day Rockefeller Center from 1801 until 1811.

View of the BOTANIC GARDEN at ELGIN in the vicinity of the CITY of NEW YORK.

Thus, from its rich roots with ties to the vision of George Washington and other important figures in American history, the United States Botanic Garden has emerged in the twenty-first century as one of the nation's foremost botanic gardens. Through partnerships with other botanic gardens, exhibits, and horticultural displays, its public outreach, conservation, and volunteer programmes, and through the scientific work it does in conjunction with the Smithsonian Institution's Department of Botany, the United States Botanic Garden ensures the nation's commitment to plant science, display and education.

The New York Botanic Garden

Gregory Long

Nathaniel Lord Britton, 1859–1934, a leading nineteenth century botanist, taxonomist and founder of the New York Botanical Garden.

The principal centre for the study of plants in pre-Revolutionary War America was Philadelphia, but by the early nineteenth century New York City had become the focal point for scholarship and higher education in plant biology. This was so because medical practice still involved extensive herbal knowledge and the New York College of Physicians and Surgeons and Columbia College were eager to foster its expansion. Dr David Hosack, 1769–1835, who taught botany on the faculty of Columbia College and maintained a large, lucrative, and socially prominent medical practice, knew that his students needed a botanic garden in order to learn from living plants. In 1801 on the site of today's Rockefeller Center, then some distance north of the settled parts of the city, he founded the Elgin Botanic Garden, forerunner of the New York Botanical Garden. Hosack invested substantial personal capital in its elegant conservatory, order beds, and a catalogue of the collection, but in 1811, when he could no longer afford to support it, the Elgin Botanic Garden ceased iterations. Hosack continued to teach, however, and two distinguished lines of botanists descended from his star student, John Torrey, 1796–1873, and Torrey's student Asa Gray, 1810–1888. Torrey and Gray collaborated on *The Flora of North America*, 1838–1843, but soon thereafter Gray left for Harvard, where he became America's most celebrated plant scientist and Charles Darwin's strongest early supporter in the United States, thereby establishing the New England branch of Hosack's educational tree.

The New York line of botanists following Hosack included other mid-nineteenth century proteges of Torrey, many of whom joined together in the 1860s to create a learned society called the Torrey Botanical Club, whose members were for the most part associated with Columbia University. It was at their meetings in the late 1880s that the idea for a new botanical garden—one with a scientific emphasis —was first formulated. The New York Botanical Garden's link to Torrey is significant in many ways, and not the least is that its research collections contain both his botanical library and herbarium.

In the period between the end of the Civil War and the beginning of the First World War, the civic leadership of New York City was intent on creating a cosmopolitan world capital. Men such as JP Morgan, 1837–1913, Andrew Carnegie, 1835–1919, John D Rockefeller, 1839–1937, and Cornelius Vanderbilt II, 1843–1899, possessed sufficient wealth to found an impressive roster of institutions emulating those they admired abroad: the American Museum of Natural History, 1869, the Metropolitan Museum of Art, 1870, the Metropolitan Opera Company, 1883, the New York Botanical Garden, 1891, the New York Zoological Society, 1895, and the New York Public Library, 1895. The New York Botanical Garden is thus part of a constellation constituting the city's cosmopolitan cultural infrastructure.

The New York Botanical Garden's professional founders were Nathaniel Lord Britton, 1859–1934, a Columbia professor of botany and geology who later distinguished himself with major publications on the trees of the northeastern United States, the plants of the West Indies, and the cactus family; and his wife, Elizabeth Gertrude

Elizabeth Gertrude Knight
Britton, 1958–1934, one of
the founders of the New York
Botanical Garden.

Knight Britton, 1858–1934, an avid and respected scholar of mosses. In 1888 the Brittons traveled to London, visited the Royal Botanic Gardens at Kew, and admired the way that the institution operated in three principal areas: as a museum of plants in a designed landscape; as a public educational programme deriving authority from the curators of its plant collections; and as an international plant exploration and research programme devoted to the study of the evolutionary history and basic biology of plants and the relationship between plants and people. Upon their return home, the Brittons launched a public campaign to establish a similar institution. Three years later the New York Botanical Garden was founded, Vanderbilt became the first president of the board, and in 1896 Britton became its first director.

The New York Botanical Garden has remained constant to a tripartite mission inspired by Kew throughout its history. However, its scientific emphasis, following that established by Britton, has differed somewhat from that of Kew in that the New York Botanical Garden has focused more on the plants of the western hemisphere. A number of distinguished figures associated with the Garden—Addison Brown, Henry Hunt Rushy, Henry A Gleason, Basset Maguire, William C Steere, Sr Arthur Cronquist, and Ghillean Prance—have perpetuated this tradition. Scientists such as Patricia and Noel Holmgren, Scott Mori, and Dennis Stevenson, and John Mickel are carrying it forward today, identifying, documenting, and publishing the plants of North and Latin America. Thus, Britton's original vision of a botanical garden oriented towards the plants of the Americas has endured for more than a century.

In the 1870s and 80s, following the example of New York's Central Park, many cities started planning parks and park systems. At the same time, social activists and urbanists in New York City began to dream of new parks in outlying parts of the growing metropolis. In 1887 John Mullaly, 1835–1911, a journalist with *The New York Herald*, published his influential book, *The New Parks Beyond the Harlem: Nearly 4,000 Acres of Free Playground for the People*, in which he described his vision for a chain of parks in the Bronx, a borough that recently had been incorporated into New York City. Extending from the old Van Cortlandt estate in the north, this system of large-scale parcels linked by wide parkways would run south to include the historic properties of the Lorillard and Bronck families (the present sites of the Botanical Garden and the Bronx Zoo, respectively) and continue eastward to Pelham Bay on Long Island Sound.

In 1884 the legislature of New York State adopted the Mullaly plan, and the resulting 'emerald necklace' remains a significant part of New York City's park system. When Britton, encouraged by fellow members of the Torrey Botanical Club, was searching for a suitable site for his American Kew, city officials offered the 250 acre Bronx Park—the central park in the new Bronx park system—as a possible site. Because of its highly picturesque terrain, its freshwater river in a rock-cut gorge, and its 50 acres of old-growth forest, Nathaniel Lord Britton fell in love with it. The New York Botanical Garden had found its home.

Calvert Vaux, 1824–1895, the designer of Central Park along with Frederick Law Olmsted, laid out the Garden's first schematic design. Unfortunately, Vaux's death interrupted the work, which was subsequently taken up by Britton himself with assistance from Samuel B Parsons, Jr, 1844–1923, and John Brinley, 1861–1946. The Olmsted Brothers, the firm originally founded by Olmsted, completed the layout of roads and pathways in the early 1920s.

From the beginning, the Garden's founders intended the collections to be comprehensive and worldwide. They dedicated propagation and exhibition space in the conservatory to tropical and desert plants and identified sites within the Garden for a deciduous arboretum, for two large-scale conifer collections, and for native plants, alpine plants, herbaceous perennials, bulbs, annuals, and roses. In the early years many plants, such as the now-mature specimen trees in the Arthur and Janet Ross Conifer Arboretum, were grown from seeds collected in the wild or from cuttings. Later, Beatrix Jones Farrand, 1872–1959, Ellen Biddle Shipman, 1869–1960, and other professional designers were retained to create gardens within the Garden for the display of new collections. In 1949 Marian Cruger Coffin, 1876–1957, designed a 15 acre landscape to house the collection of rare conifers amassed by Colonel RH Montgomery. This collection was recently restored and expanded under the supervision of Todd Forrest, the Garden's vice president for horticulture and living collections, and is now known as the Benenson Ornamental Conifers.

Today there are a million plants growing throughout the 250 acre National Historic Landmark site, representing 18,000 species or groups. The most significant collections are tropical ferns, cycads, New World succulents and palms, orchids, alpine plants, ornamental flowering trees, the deciduous trees of the northeastern United States, and the conifers of the world. These collections are exhibited within a landscape composed of venerable trees native to the site, including notable White, Red, and Black Oaks (*Quercus alba*, *Q rubra*, and *Q niger*), Tulip Trees (*Liriodendron tulipifera*), Black Gums (*Nyssa sylvatica*), and Sweet Gums (*Liquidamber styractflua*).

In addition to the living collection, the New York Botanical Garden has major research collections in its library and herbarium. Torrey's fine botanical library became the nucleus of the LuEsther T Mertz Library, which currently contains more than one million items, including books, journals, seed and nursery catalogues, architectural plans of glasshouses, scientific reprints, and photographs, and his herbarium is part of the 7.2 million plant and fungi specimens that comprise the William and Lynda Steere Herbarium.

The New York Botanical Garden recently has undertaken a comprehensive, 15 year renewal that includes strategic planning, programmatic and financial expansion, capital development, and landscape restoration. During this period, the private sector and the City and State of New York have made substantial investments in these initiatives and improvements. The educational programmes and facilities for children and adults have been expanded and the Garden has built or restored 50,000 square feet for the library and herbarium; it has added molecular research to its agenda; and it has constructed a new 28,000 square foot laboratory and 45,000 square feet of new glasshouses. In addition, it has restored many historic buildings, including the great Victorian-style conservatory and approximately 100 acres of landscape and living collections.

In spite of new developments, the New York Botanical Garden's intellectual, urbanistic, and cultural goals remain unaltered. New York City's role as an important centre for scholarship and higher education in plant biology in the nineteenth century continues in its universities and science centres, and the New York Botanical Garden is a nexus for the work of this consortium of institutions. The 1880s movement that resulted in the creation of the Bronx's system of linked parks is still alive and has become Bronx Green-Up, a New York Botanical Garden sponsored community gardening programme. Thus, within the constellation of world famous cultural institutions created during the Gilded Age, the New York Botanical Garden continues to play its role both in the life of New York City and the rest of the world.

Fairchild Tropical Botanic Garden

Mike Maunder

Located in Florida's subtropical Coral Gables, the Fairchild Tropical Botanic Garden, named after David Fairchild, 1860–1954, plant explorer for the United States Department of Agriculture was opened in 1938 to provide the residents of Miami and neighboring resort towns a glimpse of the flora of distant and exotic landscapes. As a result, visitors to the 83 acre garden today enjoy one of the world's largest collections of tropical plants. Designed by William Lyman Phillips, 1885–1966, a pioneer of tropical landscape architecture who had been a student and then partner of Frederick Law Olmsted Jr, its landscape features lakes, lagoons, and broad vistas as a frame for a combination of Floridian and non-native plant species such as palms and cycads.

Fairchild and Phillips sought to entrance the visitor with their joint vision of a tropical paradise, resembling in this regard the Renaissance princes whose gardens and *wunderkammmer*—the cabinets of curiosities that prefigure the natural history museum—displayed botanical, ethnographic, and zoological specimen of an exotic nature. Indeed, Fairchild established an ethnographic museum at the Garden and often entertained his audience with demonstrations of his skill with a South American blow-pipe.

Like Fairchild, who chronicled his discoveries and contributions to economic botany and ornamental horticulture in *The World Was My Garden Travels of a Plant Explorer*, the curators of the earliest European botanical gardens watched in awe as the world of natural history expanded with each crate of new specimens that arrived from the frontiers of exploration. Today, however, we watch via satellite television and video as the botanical world contracts, and is increasingly burned, grazed, or ploughed into oblivion. This places an enormous responsibility on all botanic gardens. The Fairchild Tropical Botanic Garden believes that in the twenty-first century it should serve as more than a series of scenes of luxuriant vegetation and collections of interesting specimens; it also must engage in issues related to the loss of global biodiversity as many species near extinction and environments undergo profound ecological collapse. The Garden therefore has made the strategic decision to support conservation in the field and in the country of origin. This has resulted in a number of changes in policy and administration. Fairchild has become an arena for interpretation and debate, and its research agenda has shifted from one that is merely academically interesting to one that addresses issues of species and habitat preservation. Its research team now works with partners in South America, the Caribbean, East Africa, and Madagascar, and it mounts exhibits that interpret the botanical diversity and environmental issues of those regions. In addition, the Garden is attempting to address problems in its own community. In the 1930s, before Miami had undergone rampant expansion, David Fairchild, who deplored the effects modernisation was having on the native cultures with which he was familiar through his far-flung explorations, predicted its deleterious consequences at home. Today Miami, like other large cities, has an increasing number of citizens whose lives are divorced from the natural world; they have not seen a growing pineapple or banana plant, never stood in the shade of a native woodland canopy, or watched a hummingbird. Their plight, which may be described as 'bioilliteracy', is not one that is restricted to poor urban neighborhoods.

Seen in this light, parrots provide a useful parable. According to legend, the eighteenth century German scientist Baron Alexander von Humboldt, 1769–1859 was travelling on the Orinoco River in Venezuela when he encountered a Carib

Snake With Ipomoea Ochracea, a hand-coloured engraving from Albertus Seba's *Locupletissimi rerum naturalium thesauri.*

Palm tree grove at the
Fairchild Tropical Botanic Garden.

Indian tribe. Humboldt noticed that their pet parrots were speaking a dialect different from that of their owners. The Indians explained that the birds had belonged to the Maypure tribe, whom they recently had exterminated during a tribal conflict. The parrots were the last remaining speakers of Maypure, the unwitting and ornamental custodians of a language they could neither understand nor conserve. The implications for botanic gardens that are perceived primarily as places of exotic plant display are clear: how do they avoid becoming like Humboldt's parrots, squawking an incomprehensible rhetoric about conserving almost extinct species when what their visitors experience is a vision of paradise?

This dilemma was the focus of recent strategic revisions at Fairchild. Following discussions with staff, volunteers, donors, and board members, Fairchild determined to be more than a pretty parrot cage, a garden intended simply for viewing tropical vegetation. First, it sought to define the role of a botanic garden in a city where most people were born elsewhere and where the prevailing cultural influences are from Latin America and the Caribbean. Secondly, it sought to identify how a botanic garden tackles environmental education and stewardship in a city with as many poor and culturally diverse residents as Miami. Thirdly, recognising that species and habitats cannot be saved within the confines of a botanic garden, it sought to understand how the Garden could truly play an effective role in preventing the further extinction of plant species elsewhere.

Defining its duty to confront the biodiversity crisis and bioilliteracy dilemma does not mean that Fairchild should neglect its original mission to provide visitors with an experience of delight, wonder, and fascination. After all, how many people

come to Fairchild to seek an understanding of the perianth structure of the *Melastomataceae* or to discuss the implications of climate change? They do come to enjoy shaded walks and to admire orchids, hibiscus, tropical water lilies, and the occasional flowering of the Giant Arum. Garden officials have therefore worked to increase the number and abundance of flowers, to create a sense of welcome, and to host art and music events in the Garden. These efforts have broadened the attraction of the Garden to a wider range of communities and cultures.

At the same time, festivals centred around orchids, butterflies, and mangos serve as a means of promoting an understanding of ecological issues and concepts. In addition, to develop a sense of individual responsibility for all landscapes, whether or not they are endangered, the Fairchild Challenge, an environmental education outreach programme for middle schools and high schools was created. Schools participate in such diverse course options as the fine arts, website design, gardening, science, habitat restoration, community service, creative writing, photography, environmental debate, and ethnobotany. In 2006 an estimated 16,500 students from 63 schools took part in the Fairchild Challenge.

Every botanic garden is a combination of historical and contemporary influences. While Fairchild is a relatively new garden, it has undergone a series of dramatic cultural changes. It has listened to Baron von Humboldt's parrots and taken note. Like many other botanic gardens, it is developing a new institutional culture that is both socially relevant and culturally audacious. Although the mission of the Fairchild Tropical Botanic Garden is increasingly focused on combating species extinction and overcoming bioilliteracy, the twenty-first century garden—the public face for Fairchild's mission—is still true to David Fairchild's original vision of a tropical *wunderkammer*, a garden of revelation and enchantment.

David Fairchild, in 1940, wading the Kasareota River, Indonesia in search of exotic plants. Courtesy of Fairchild Tropical Botanic Garden Archive.

JAPAN
BOTANIC GARDENS OF TOYAMA

There has been a strong appreciation of all forms of plant life throughout Japanese history, and this is still clearly seen in the culture of today, with more than 60 botanic gardens, of varying size, to be found scattered across the stratovolcanic archipelago. Japan is a biodiversity hotspot, which is not surprising when one considers the great variations in climate that can be found there, from the tropical islands of the deep south to the far more temperate climes of the north. These variations are further highlighted by the mountainous nature of the Japanese archipelago. This has naturally led to the evolution of a great number of plant species endemic to Japan.

The Botanic Gardens of Toyama were begun in 1989, with an outdoor exhibition garden opening in October, 1993. After the construction of a number of display greenhouses and a sun-roofed hall, the Gardens were fully opened to the public in April 1996. One more addition, the Yunnann greenhouse, was added in February 2000. The Botanic Gardens of Toyama are the first general botanic gardens of there kind on the Japan Sea coast. It is equipped with all the necessary facilities to cultivate, preserve, investigate, and research plants.

The Botanic Gardens at Toyama offer a tropical plant house, orchid house, tropical fruit house and Alpine plant house. There

is also a garden dedicated to Japanese tree peony, plants native to Yunnann, fragrant plants, a promenade of flowers, a clematis garden, and an extensive selection of coastal plants and other species from low-lying and marshy plains. There are more trees on show in the form of forests of chestnuts and the magnificent Japanese Oak.

The Gardens now cover nearly 62 acres and display over 5,000 native and foreign plants. There has been a great attention to detail in all aspects of planning the site, with the delightful flower promenade running right through the centre of the Garden's museum. On the northern side of this lavish promenade the World Plant zone can be found, with an area of Japanese plants placed at the opposite end. The glass atrium at Toyama accommodates a greenhouse of rainforest plants, fruits and orchids, and a cold room, which displays the Garden's selection of alpine flora and fauna.

tag: *Hylotelephium sieboldii,* an endangered species of plant of which is distributed only in Toyama. opposite top left: Historical drawing of *Flora Japonica.* opposite top right: 'Yukituri', structures traditionally used by the Gardens to protect Evergreen Pines from snow fall. opposite bottom: A collection of plants from tropical and subtropical areas of the world exhibited in the Tropical Rain Forest House. above right: *Prunus yedoensis* in full bloom reflected in the Gardens' pond. below: View from Cherry Walk across the pond to the Gardens' Greenhouses.

LITHUANIA
VILNIUS UNIVERSITY BOTANIC GARDENS

The Vilnius University Botanical Gardens have a long history. It was founded in 1781 by Professor J E Gilbert, on the grounds of Vilnius University, and has since changed its location twice. At one point the original gardens contained 6,565 species, though many of these perished when the University was closed down in 1832. In 1919 new Botanic Gardens were opened in Vingis park, within the Neris river valley. These later became home to the University's Department of Plant Systematic and Geography, with today's main Botanic Gardens, begun in 1974, taking up residence in Kairenai. This site opened to the public in 2000, and every year since has hosted the International Day of Biodiversity and a highly acclaimed exhibition entitled *Nature and Art*.

The modern Gardens cover 491 acres, offering over 9,000 varieties of plant and flora. It lies some 150 metres above sea level, with temperatures varying between minus 35 degrees in the depths of winter and 35 degrees during the summer months. There are five departments within the Gardens including dendrology, floriculture, pomology, plant genetics and plant systematic and geography. The Gardens also house The Laboratory of Plant Physiology and Isolated Issue Cultures.

Main research trends at Vilnius include: the introduction and investigation of ornamental herbaceous and ligneous plants, the introduction and selection of small fruit plants, mutagenicity and genetic instability of plants (particularly in barley and in beans), genetic collections, and in-vitro micropropagation.

The University's Department of Dendrology is the largest research body at the Gardens, incorporating the old Kairenai Park, the Forest Park and the Arboretum. The project, based on geographical and ecological principles, aims to include the floral zones of Europe, the Mediterranean, the Caucasus, Siberia, North America, the Far East, China, Japan and Central Asia. Research involves the investigation of woody plants in their introduction to a new environment—their hardiness and bio-ecological characteristics—and growing technology. There are over 2,500 species in the dendrology

collection, comprising 229 genera, 79 families and about 1,200 varieties of wild plants.

The Department of Plant Systematic was founded in 1993, and stores over 3,000 varieties of plant life. The bulk of these are from the greenhouse, bulb and plant systematic collections. The reproductive process of over 400 cultivers of tulip bulbs has been studied here.

The Department of Floriculture was founded in 1992 and since then has gathered over 3,200 plants from more than 80 families and 253 genera. Most importantly, this is the first time that a unique collection of flowers has been created by local breeders in Lithuania. Special nurseries have also been created for the conservation of the gene bank of these flowers. The floriculture department houses perhaps the most visually pleasing area of the Gardens; an abundance of flowers circle the ponds, with each bed being arranged by season. Three of the ponds are surrounded by The Flower Valley, displaying field flowers which blossom from spring to autumn, and belong to different bio-ecological groups.

Some of the most interesting research taking place at the Vilnius University Botanic Gardens occurs in the Department of Plant Genetics. One prominent area here involves the investigation of barley mutants, with two particular strains, that of the Branded Ear and Tweaky Spike, being examined most closely.

Main research areas at the Department of Pomology include the selection and introduction of berry plants. The current collection contains plants of 14 genera, 105 species and 747 cultivars from 27 countries.

The Botanic Gardens at Vilnius is an exciting centre of plant research, which also offers the chance to enjoy a wide range of plants, flowers and fauna in a stunningly beautiful northern setting.

tag: Berries from a Sorbus plant. photograph: S Zilinskaite. opposite top: The Botanic Gardens has moved four times during its lifetime. This historial depiction shows the Gardens while located in Sereikiskes, from 1799–1842. photograph: K Racinskas. opposite bottom: A nineteenth century fountain in the Gardens. photograph: A Ufartas. above: Collection of Dahlia genus plants in the Department of Floriculture. photograph: D Ryliskis.

Situated some 17 kilometers north of the village of Toliara, and two kilometers north of the Tropic of Capricorn, the Antsokay Arboretum has been described as a sightseeing must by those holidaying in the region. It was founded in 1980 through the work of Swiss amateur botanist, Hermann Petignant. It had humble beginnings, with Petignant first acquiring a number of poor plots—typical of the arid nature of this part of the island, which were covered by limestone and red low-calcium sand. Antsokay, the nearby hamlet from which the Arboretum takes its name, has ample reserves of limestone, which can be used for the production of quick lime.

One of Pertignan's first objectives, coming from concerns about constant deforestation, bush fires and looters in the area, was to reproduce the most threatened species of plant, whether by seeding, cutting or a process of transplantation. Through this rounding up of local species—with Pertignan and his team making constant expeditions into the bush and unearthing a number of new specimens—plants could be identified and studied. The grounds of the arboretum now cover over 98 acres, and are surrounded by protective vegetal hedges.

The Antsokay Arboretum works in close collaboration with many institutions dedicated to the preservation of the environment, though it does take protection of plants from the south-west of Madagascar as its priority. Other participating bodies include the Royal Botanic Gardens, Kew, the WWF, the CEPF and the Fairchild Tropical Botanic Garden.

After 25 years of a great deal of hard work the Antsokay Arboretum offers a densely kept selection of flora to the visitor, presenting a typical spiny thicket at a sub-arid stage with the canopy of a number of Baobab trees noticeable above the smaller plants. There is a botanic garden of four hectares open to the public, which comprises a selection of plants from the south of the island, and over 900 species of plant on display amongst which can be found a number of endangered species. 90 per cent of the collection is endemic to the region and 80 per cent has medicinal value.

The Antsokay Arboretum have three main objectives including the maintenance of nurturing of a collection living and preserved plants from the south-west of Madagascar; implementing conservation and botanical research programmes; and promoting public awareness and knowledge of plants and the importance of their conservation.

Preserved plant specimens are kept in the herbarium at Antsokay. The collection here comprises species either collected by staff members or donated by or exchanged with partners and other herbariums in Madagascar. These specimens can be of scientific value for up to 100 years and data taken from them is essential for any decision, making on conservation action. The herbarium collection also represents a comprehensive report on biodiversity through the ages.

tag: *Alluaudiopsis marnieriana,* located in the Display Area of the Arboretum. opposite top: The Arboretum's main entrance. opposite bottom left: Fony Baobab, (*Adansonia rubrostipa*), located in the Aboretum's Natural Area. opposite bottom right: Palms (*Hyphaene coriacea*), located in the Display Area of the Arboretum.

MEXICO
BOTANIC GARDENS OF THE AUTONOMOUS NATIONAL UNIVERSITY OF BOTANY, MEXICO CITY

Mexican gardens were famous before the arrival of the Spanish. The indigenous population loved and respected the natural world around them, maintaining great gardens comprised of flora from far away places and plants which were useful to them for their healing properties. Two of the most notable of these pre-Hispanic gardens were those at Texcoco and Huaxtepec, the latter being founded by Moctezuma.

Due to its geographical location Mexico offers some of the greatest floral diversity in the world, and the Gardens at the National University gives one the chance to see a large part of it. They are centres of biological investigation, teaching and support to various educational programmes. They also play an important role in the conservation of flora, housing in their collections endemic, rare and endangered plants.

Work on the Gardens commenced in 1958 when the Biology Institute of the Universidad Autonoma moved from the Lake House of Chapultepec to Ciudad Universitaria, where it was necessary to build installations for the maintenance of plants chosen for study. The esteemed Hispano-Mexican botanist Dr Faustino Miranda designed a garden dedicated to the cultivation and conservation of Mexican flora while Dr Efren del Pozo, then secretary general of the Universidad Nacional Autonoma de Mexico (UNAM), oversaw the creation of a programme for the propagation of ornamental plants. With the fusion of both of these projects the Botanic Gardens of the UNAM was founded on 1 January 1959.

The mission of the Botanic Gardens of the UNAM is to investigate the use, management and cultural value of Mexican flora, focusing in particular on *Agaucaeae*, *Arecaceae*, *Cactaceae* and

Orchidaceae—both their historical value, and their present and future state. It is of fundamental importance to the Gardens to work actively toward the conservation of all plant resources; to detect and evaluate the state of endemic, rare or endangered species, to investigate conditions required for their growth with the aim of maintaining these collections, and to spread and make available all knowledge gained. Many studies require periodic and constant observations, and these are made easier by having a collection of living plants ready to hand. Another objective of the Gardens is the teaching and spreading of botany as an academic discipline. There is a constant flow of visitors to the Gardens, all of whom come in search of information on a wide range of botanical themes. In response to such interest the institute puts on guided tours, courses, conferences and workshops amongst other services.

The UNAM has created three main routes for the visitor through the grounds that take 'scents', 'curiosities' and 'forms' as their starting point with the intention of giving as diverse and complete a look as possible at all the University's collections to those interested in botanical study. The aim is for the visitor is to learn and enjoy themselves while exploring the plants of the Gardens, to search for, observe and finally find a common pattern in the series of exhibits which form each particular route. Those who come to the Gardens are also encouraged to visit at other times of the year, to be able to observe the changing behaviour of plants through different seasons. These routes are an excellent way to develop one's personal interest in plant life, while supporting the University's investigations and dispersal of its flora collection.

tag: *Echinocactus,* one of the Gardens' many cacti. previous pages bottom: A panoramic view of the National Collection of Agavaceae. previous pages top: The Gardens' National Collection of Agavaceae. previous pages bottom right: The Aquatic Plants Collection. left: Leaves of the succulent Agave plant. right: *Beaucarnea gracilis,* an endangered species.

Monte Carlo is one of the most densely populated cities in the world, with a population of 30,000 squeezed into an area of just 1.95 kilometres. As such, the Exotic Garden of Monaco provides the area with some much-needed breathing space.

The Garden is positioned on a craggy cliff, called "The Observatory" after the small astronomical observatory once located there several hundred metres above sea level. Opened in 1933, its collection is most famous for its abundance of succulent plants. Monaco is known for its warm climate, and provides a good home to this family of plants, of which the cacti is the best-known variety. Succulent plants have developed several types of features to cope with dry climates, the most spectacular of which is the presence of a hypertrophied organ (usually in the form of a stem) that stores reserves of water, and to further retain water, their leaves have been replaced by spines and thorns. One of the most striking aspects of this botanic garden is the sheer quantity of cacti sprawling up the cliffs. During the summer, there is a shower of colour when many cactus species produce large colourful blooms. During the winter, aloes and African Crassula bloom.

The plant species represented in this Garden come from arid zones all over the world, including the south-western United States, Mexico,

and Central and South America, southern and eastern Africa and the Arabian peninsula. Despite their unusual shapes, these are plants like any others that regularly blossom to propagate.

The most interesting feature of this botanic garden is not really botanical at all. At the base of the Exotic Garden's cliff, at 100 metres in elevation, there is an opening to an underground cavity, which has been fitted out for tourist visits. The calcareous rock was progressively hollowed out by water rich in carbon dioxide and is full of caverns complete with amazing concretions: stalactites, stalagmites, curtains, columns and spaghetti-like helictites.

tag: Cactus (*Cactus mammillaria*). previous page left: View of the Garden's extensive Cactus Collection. previous page bottom: *Lampranthus haworthii,* which originates from South Africa. opposite: Main promenade of the Garden. below: View of one of the many pathways leading through the Garden.

MOROCCO
MAJORELLE GARDEN, MARRAKECH

The Majorelle Garden in Marrakech probably lays claim to the most stylish patronage a botanic garden could have. In 1980, the Garden was bought by fashion designer Yves Saint Laurent and his partner Pierre Bergé, who set about redeveloping the site in its entirety, making what it is today. There is indeed, a certain Parisian glamour to this Garden that it shares with no other.

The Garden was originally owned and designed by French oil painter Jacques Majorelle, who settled in Morocco in 1919, in an attempt to cure his tuberculosis. In the bustling souks of Marrakech he discovered a love of simple forms and bold Fauvist colours that he expressed in his art. He also discovered an avid interest in amateur botany, and shortly after his arrival, he purchased some land on the outskirts of the medina, and set about transforming the grounds. What started out as a few potted plants on his terrace, slowly evolved into a small paradise of tropical yuccas, jasmine, bougainvillaea and cacti. In 1931, Majorelle asked architect Paul Sinoir to design a studio for him in the style of a typical Moroccan palace, and in 1947 the magnificent grounds were at last opened to the public. A mere 15 years later, however, the gates were once again shut, after Majorelle was injured in a car crash and suddenly left for France. He never returned to his house or garden in Marrakech, and the site subsequently went through a period of deterioration that lasted almost 20 years. Yves Saint Laurent began its renovation in 1980, and it took a further 20 years to complete the process, with the doors opening at last in January 2001, and the Garden restored to all its original glory.

This botanic garden is categorised by a lushness and luxury that one would not usually expect from an arid country like Morocco. It is an immensely decadent garden—in scale, height and colour—and its French ancestry is very much apparent. Every surface, from the studio's exterior walls to the colossal flowerpots, is painted a vivid cobalt blue, setting a stark contrast to the red clay walls of the Marrakech medina. The vegetation is dense and spectacularly green: Majorelle would bring plants back from his trips to Europe and Northern and sub-Saharan Africa and the collection now contains wonderful examples of dragon tree, cypress, jacaranda and aloe.

As well as the French influences, an Islamic aesthetic is also very much apparent. Water is historically and culturally a common feature in any Arabic garden and this is reflected in the square pools of water, filled with waterlillies and lotuses, as well as the numerous streams that flow through the Garden. The Islamic Art Museum is now housed in what was once Majorelle's studio in the middle of the grounds, exhibiting Islamic and Berber artefacts, as well as a number of Majorelle's original canvasses.

tag: One of the many cacti cultivated by the Garden. opposite top: One of the Garden's characteristically brightly coloured buildings, surrounded by cacti and palms. opposite bottom: An Islamic architectural detail from one of the Garden's brightly coloured buildings. top: Waterlilies and other water-based plants nestle densely in the Garden's pond. bottom: The Garden's tiled fountain, which echoes the tradition of Islamic architecture and the vibrant colours of the Garden.

THE NETHERLANDS
LEIDEN UNIVERSITY BOTANIC GARDEN

In 1587 the University of Leiden requested the burgomasters of Leiden permission to establish a Hortus academicus behind the Academy building, for the benefit of the medicine students. Permission was granted in 1590, and the famous botanist Carolus Clusius, 1526–1609, was appointed prefect. Clusius' knowledge, reputation and international contacts allowed him to set up a very extensive plant collection, and he urged the Dutch East India Company to assist him in collecting plants and dried plant specimens. The original garden set up by Clusius was nothing but a tiny plot of land—about 35 by 40 metres—but on it he managed to grow more than 1,000 different plants.

The collecting of tropical (from the Indies) and sub-tropical (from the Cape Colony) plants was continued under Clusius' successors.

In particular, Herman Boerhaave (1668–1738, prefect from 1709–1730), contributed greatly to the fame of the Hortus with his efforts to collect new plants and specimens, and with his publications, such as a catalogue of the plants then to be found in the Garden.

Another major contribution to the collections was made by Philipp Franz von Siebold, a German physician who was employed for a period in Japan, during which time he collected many dried and living plants from all over the country and sent them to Leiden.

The first greenhouses appeared in the Hortus in the second half of the seventeenth century, and the monumental orangery was built between 1740 and 1744. From its original plan the Hortus was progressively expanded until 1817. In 1857 a part of the grounds was used for building the new Leiden Observatory.

Today, the rich history of Hortus Botanicus Leiden, means that the Garden contains some fascinating specimens. For example, the old Golden Chain (*Laburnum anagyroides*), planted in 1601, the Tulip tree (*Liriodendron tulipifera*) from 1682 and the Ginkgo (*Ginkgo biloba*) from 1785. From a scientific point of view, the Leiden Hortus is known for its collections of Asian Araceae (among which the *Amorphophallus titanum*), hoya, dischidia, Asian orchids and ferns. It's beautiful greenhouses play host to a vast selection of tropical species, including the Victoria amazonica, whilst the recently renovated Orangery holds a large collection of sub-tropical plants. Undoubtedly one of the most interesting sections is the Clusius Garden. Founded in 1931 and renovated in 1990, this is an exact replication of Clusius' original garden, based on a plant list dating from the end of the sixteenth century.

tag: The *Odontoglossum wyattianum* orchid, which originates in Peru. opposite: The Greenhouse, which shelters many tropical and warm climate plants from around the world. top: The systematic garden in front of a townhouse type building. bottom: The Orangery in spring.

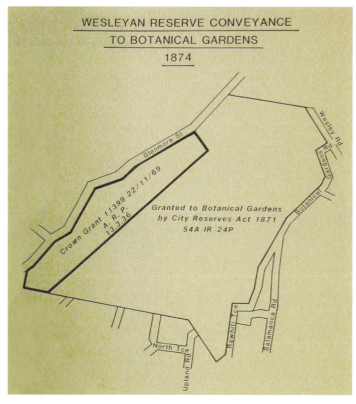

The dense, rich forests of New Zealand that confronted the European settlers who arrived in country in the early nineteenth century, must have seemed simultaneously intimidatingly foreign and full of the promise of potential wealth. In 1839, the New Zealand Company instructed surveyor William Mein Smith to set aside land for a town belt, a reserve for a botanic garden and a park, providing the settlers with an important conceptual and cultural link to Europe. In 1844, a 13 acre strip of this land was placed under the trusteeship of the Wellington Horticultural Society for a Botanic Garden Reserve. Due to intense logging, most of that strip had been cleared of its native vegetation, and in the 1960s, the head of the Botanic Garden Board, Sir James Hector, managed to persuade the government to acquire the adjoining 54 acre Wesleyan Reserve where some native forest remained. The merging of these two areas ensured the survival of a fragile remnant of native bush, as well as providing the land for a cultivated park.

Wellington Botanic Garden was officially established in 1868 and managed by the New Zealand Institute, which planted the major conifer species that can be seen today. Wellington City Council has managed the Garden since 1891, developing the award-winning

Lady Norwood Rose Garden in 1950, the Begonia House, 1960, and the Treehouse Visitor Centre, 1991.

The Garden today covers 62 acres and is a unique mixture of protected native forest, conifer plantings and plant collections with major seasonal floral bedding displays. Exotic conifers are a significant component of the Wellington Botanic Garden. Many of the 46 species of pines and conifers growing in the Garden are a legacy of the early trials of the 1870s when many overseas species were propagated, grown and then assessed for forestry potential. The Monterey pine, *Pinus radiata*, was one of the first to prove its vigour. Those growing today on Druid Hill and Magpie Spur are some of the oldest exotic trees in New Zealand.

Sheltered gullies in the Garden contain remnants of native forest that clothed the landscape before European settlement. The oldest tree in the Garden, a gnarled hinau in the Stable Gully remnant, is over 200 years old. Sadly, the original forest giants, including the Giant Rimu and Totara, are long gone. However, there are plans to reintroduce these missing components of the native flora to allow future generations of visitors to enjoy these magnificent sights once more.

tag: A selection of cacti in the succulent garden. opposite top: Flowers and dappled light and shade enhance the reflective mood of the Bolton Street Memorial, within the Garden. opposite bottom left: Aerial view over ferns to the Garden's picturesque duck pond. opposite bottom right: Historical map showing the garden's inception in 1874. left: Aerial view over ferns to the Garden's picturesque duck pond.

NORWAY
TROMSØ UNIVERSITY MUSEUM ARCTIC-ALPINE BOTANIC GARDEN

Located in Tromsø, Norway, the Arctic-Alpine Botanic Garden is the northern-most botanic garden in the world, and features an impressive display of arctic and alpine plants from all over the northern hemisphere. Situated on the grounds of an old farm in Southern Breivika, the Garden's land was donated by Ms Hansine Hansen in 1938. The remaining parts of the farm were given to the County of Troms upon her death in 1947, and large parts of the University of Tromsø, as well as Breivika High School, are also located on the property.

The location of this Garden is notable in a number of ways; it offers striking views, illustrating the extreme arctic climate, however, thanks to a branch of the Gulf Stream that sweeps up the northern coast of Norway, the climate in Tromsø is relatively moderate, with mild winters and cool summers. These climatic conditions allow the Garden's curators to employ a technique called Windbreak Planting, in which plants and shrubs are distributed in relation to the wind force and the seasonal changes, allowing for more growth control in tune with the landscape.

Plant collections in the Garden are arranged according to geographical or botanical themes. The herbal garden, partly constructed on the site of ancient monastery gardens, contains a wide selection of plants within the major sections of medicinal herbs, spices, poisonous plants, plants for dyeing, aromatic plants, aphrodisiacs and plants connected to witchcraft in earlier centuries.

It is a point of interest that the world's northern-most botanic garden should have a collection of plants coming from the southern-most regions. The main focus is on South America because botanists at the University of Tromsø have studied the flora of southern Chile. Further exotica include Dainty Alpine Columbines, Low-Growing Stonecrops, Blue Monks Heads and the Woolly Willow, a threatened species from the Svalbard archipelago off the northern coast of Norway.

The period of herbaceous growth in this Garden is relatively short due to the limited sunlight that the plants are exposed to from October until April. Between May and July, however, the sun is continuously above the horizon, allowing for a brief period of intensive growth. The opening season for this Garden is therefore only between the months of May and October.

tag: *Primula bhutanica,* a rare species originating from Bhutan. left: *Erythronium sibiricum,* a Siberian species which self-sows. opposite top left: An unidentified Calceolaria species from Chile. opposite top right: The Mossy Saxifrage (*Saxifraga*) Collection. opposite bottom left: Dianthus superbus, a species growing in the north-east arctic corner of mainland Norway. opposite bottom right: Saxifraga matta-florida, a Himalayan Porphyrion species.

OMAN
SULTAN QABOOS UNIVERSITY BOTANIC GARDEN

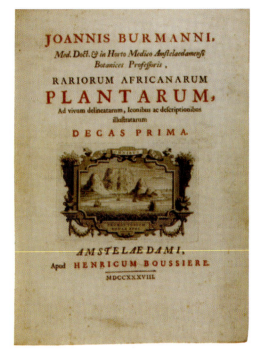

The Sultan Qaboos University Botanic Garden, founded in 1989, is a landscaped area of 11 acres on the site of the Sultan Qaboos University, in Muscat, Oman. It is the only botanic garden in the Arabian peninsula, and the only garden in the world specialising in Arabian plants. Despite the dry desert climate of inland Oman, and the humid coastal environment, Oman is home to over 1,200 species of plantlife, with 56 of these species unique to the Sultanate. Many of these endemic examples find prominent display in the Oman Plant Collection at the Garden.

The Garden is divided into themed sections, each of which is designed to create a specific environment for a given species. At the centre of the Garden is an Aquatic Garden, featuring a broad variety of pond and swamp plants as well as a falaj—an Omani irrigation system consisting of a well and subterranean irrigation channels.

The Ibn Sina Herbal Garden focuses on the plant species from Arabia that are most often used for medicinal purposes. It is important to remember that over 60 per cent of the world's medicines come from plants so collections like these are therefore vital in the ongoing investigation of the healing properties of flora. This zone also accommodates the Functional-Morphology Beds, which house plant species that illustrate various pollination and seed dispersal methods. There are also displays of typical adaptations of desert plants.

Tree species from other areas are well represented in the Arboretum, Fiscus Collection and African Collection, with large numbers of Omani trees also providing a native presence.

tag: *Delonix elata*, a flowering tree which is native to Madagascar and Africa. above: Johannes Burman's *Rarioum Africanum Plantarum*, 1739, an illustrated record of rare and medicinal plants originating from Africa. middle: A view of the Garden from the rockery and pool. below: A Frankincense Tree, a species which is often used in a variety of medicinal and cosmetic products.

PAKISTAN
LAHORE BOTANIC GARDEN

The city of Lahore is known within Pakistan as the 'City of Baghs' or gardens, thanks to its four botanic gardens as well as its many smaller parks and landscaped areas. Dating back to the Mughals, who ruled the area from the fifteenth to eighteenth centuries, the traditions of ornamental and instructive landscaped gardens are deeply rooted in the city's history.

The Botanic Garden at the University of Punjab is one of the key gardens in Lahore, which shares its space with the Qaid-e-Azam campus of the University of Punjab. Although the Garden was only established in 1964, as part of the Department of Botany at the University, its aesthetic references reach far back into the eighteenth century and beyond. Taking inspiration from Mughal tradtions, the Garden is beautifully and geometrically arranged—designed to provide a cool and sensual respite from the harsh sun. Great emphasis is placed on fragrance and the presence of water, and the 30 acre plot features a canal, and smaller ponds.

Currently, the Botanic Garden is home to many hundreds of species which are being documented and added to on a regular basis by students and professors at the University. It includes a number of wild flora as well as exotic and rare species, and the recent addition of a herbal plot. It also features a palm grove, the structural planting of which makes the most of the striking, spiky foliage by placing the palms around a lush, manicured lawn.

Rising awareness and concerns regarding the ecology of Pakistan, means that the Garden has taken on an even more academic focus, with extensive research being carried out into local vegetation. Experimental fields and plant breeding enclosures facilitate the conservation of germplasms of endangered and economically significant plant species.

tag: The *Cycas circinalis* palm. Though the plant contains toxins, the seeds can still be ground into Sago flour after careful processing. top: An eighteenth century Rajput miniature which demonstrates the long tradition of Moghul pleasure gardens. bottom: The *Cassia corymbosa* bush, a member of the pod-producing Senna family, is covered with buttercup-like blossom during flowering season.

WARSAW AGRICULTURAL UNIVERSITY BOTANIC GARDEN

The Warsaw University Botanic Garden is not only Poland's oldest Botanic Garden, it is also one of its smallest and most resilient. The Garden contains over 5,000 plant species withing its 12.4 acres of land. One of the plants most widely associated with the Garden is the ancient Maidenhair Tree (*Ginko Baloba*), the only sole survivor of its species with no living relatives. Sometimes known as a 'living fossil', the Maidenhair tree is a fitting symbolic metaphor for this garden's turbulent 175-year history.

The Garden was initially titled the Royal Garden, when the land on which it was based was given to the Polish Government by the Russian Tsar, Alexander I, in 1818. Just six years later the Garden housed 11,000 species, including 1,000 native Polish flora. During the November Insurrection of 1830—an armed rebellion against Russia's rule in Poland—over two thirds of the Garden was destroyed. The area then became a stage for political unrest, with mismanagement by several directors only adding to the Garden's demise.

In 1916, the Garden became part of Warsaw University, when its director, Professor Zygmunt Woycicki, pragmatically lead the Garden's restoration efforts. Professor Bolesaw Hryniewiecki, a prominent plant taxonomist and geographer, superceded Woycicki as director of the garden in 1919 and continued his good work.

Despite setbacks caused by the Second World War. and financial difficulties, which forced the Garden to lose its scientific status in 1987, the Garden's current director, Dr Hanna Werblan-Jakubiec has sucessfully returned the Garden to its natural beauty. Today it comprises 12 sections, and is home to an ever-expanding collection of plant life.

At present the largest section of the Garden is the Plant Systematics Section—designed to reflect the taxonomical classification created by Adolf Engler in the late nineteenth century—home to over 700 species of plants. Visitors can walk below the flowerbeds to the pretty Climbers' Garden whose housed plants stretch upwards above shrubs and trees, to the sunlight above. In the nearby Rose Garden Warszawa, Kutno and Chopin can be found, roses that are native to Poland. In the Garden's Ornamental Section, visitors can observe the ornamental flowers used in many architectural complexes, palaces and public spaces. Each year a new ornamental flowerbed is planted in the centre yard with a different nineteenth century inspired design. Ancient cereals, herbs and species, and other plants important in every-day life, grow in the economic section. There is also a garden dedicated to medicinal plants.

Despite these thriving outdoor gardens, a number of succulent tropical plants also live inside the Garden's three greenhouses. The Tropical Greenhouse and Palmhouse are home to palm trees and shrubs, some of which are 40 years old. The Misty House contains epiphytes from the pineapple, the arum and the orchid families. There is also a subtropical greenhouse that is home to tropical water lilies and ferns and economic plants like ginger and coffee. The Garden's Succulent House hosts Mediterranean plants. Inside is one of the last links to the Garden's early beginnings: an eighteenth century Trebhaus and a European fan palm that was planted in there before the Second World War.

tag: Leaves of the Maidenhair Tree (*Ginkgo biloba*), one of the oldest trees in the Garden. opposite top: Interior view of the Garden's new Palm House. photograph: Marcin Zych. opposite bottom left: View of a meteorological booth, situated in the Polish Lowland Flora Section. In the background are flowering Rhododendrons, part of The Ornamental Plants Section. opposite bottom right: View of the Rose Garden and the Warsaw University Astronomical Observatory in winter. photograph: Marcin Zych.

The Gardens at the University of Coimbra, 200 kilometres north of Lisbon, were established in the eighteenth century by the Marquis da Pombal, and integrated into the Natural History Museum, which he also founded. Having been expanded throughout its existence, the Garden, widely believed to be one of the most beautiful in Europe, now comprises 32 acres. The grounds can be roughly separated into two parts; a formal garden on the lower level of the University, laid out in terraces, and an Arboretum on a raised plateau. The lower garden, known as the Quadrado Central or Central Square, belongs to a tradition of landscape design established in the affluent European estates of the late eighteenth century. The formality of the terraced garden, with its taxonomically grouped beds allows for an understanding of horticulture through separation, and conveniently serves the interests of the students of botany at Coimbra, as well as similar institutions worldwide with whom plants are exchanged.

The Arboretum on the other hand, is noteworthy for its selection of bamboos and exotic trees as well as its renowned collection of eucalyptus species. It is home to extensive wildlife, including the Brown Squirrel, (*Scuirus vulgaris*), which was re-introduced in 1994, and has now bred into a thriving population. At the centre of the Arboretum are the university's original botany library, herbarium, museum and laboratories. The herbarium is of particular importance, housing over one million herb specimens from all over the world.

tag: The Garden's main promenade. opposite top: The Garden's Large Greenhouse. opposite bottom left: View of the Garden's Central Square and fountain. opposite bottom right: The São Bento Chapel surrounded by the Garden's Bamboo Forest. right: The D Maria I Gate, constructed in 1791. all photographs: Hugo Marques.

ROMANIA
BOTANICAL GARDEN DIMITRIE BRANDZA

The Botanical Garden Dimitrie Brandza at the University of Bucharest is a teaching, education and research centre, focused on understanding, documenting and conserving plant diversity. Founded in 1860, during the reign of Alexandru Ioan Cuza, the Garden was later renamed in memory of Professor Dimitrie Brandza, a distinguished Romanian botanist, who organised and managed this Garden between 1884 and 1895. At present, the Garden covers an area of 43.2 acres.

More than 5,000 plant species, from within Romania and from all over the world, are divided into sectors, and organised according to genus. From the scientific point of view, special attention is given to the Rare Plant sector, dedicated to ex situ conservation of nationally endemic and rare plants, such as *Salvia transsilvanica*, *Galanthus elwesii* and *Crocus biflorus*. Endangered species are also maintained in this garden, making it a valuable resource. To this end, but meriting their own sector, is the Garden's significant collection of Dobrogea Flora, which features the rare flora only found in the south-eastern part of Romania. A sector dedicated to Mediterranean Flora impressively displays more than 200 taxa, some of which are variations on familiar Romanian species.

By contrast to these little enclaves, the Italian Garden covers a large space in the central part of the Garden, between tree-lined alleys dating back to the nineteenth century. The area also contains a small pond, which provides shelter for numerous species of native wildlife.

The Botanical Garden Dimitrie Brandza places a strong emphasis on public awareness and education concerning the natural world. Enrichment programmes are run for school children from kindergarten up, teaching a range of subjects dealing with nature and environmental preservation. The Garden also boasts a Botanical Museum, a General Herbarium and a Botanical Library. In 2000, this Garden founded The Association of the Romanian Botanic Gardens, networking all nine botanic gardens from Romania. The Garden's scientific publications *Acta Horti Botanici Bucurestiensis* and *Delectus Seminum* are further methods of disseminating information about research activities and seed collections.

tag: A Sumar Plant (*Dicentra spectabilis*), located in the Garden's Decorative Section. below: The Botanical Museum. opposite top: The Fruit Collection of the Botanical Museum. opposite bottom left: The Garden in winter. opposite bottom right: The Fern Section of the Garden's Greenhouse. all photographs: Anca Sarbu.

The Botanic Garden of Rostov State University was originally commissioned in 1927, to focus on the study of native flora and botanical life. It started off as a strip of 182 acres along the Temernik River, and over the 78 years of its existence, it has grown in both size and reputation. Affiliated to the State University of Rostov, the Garden is located in the north-west of this southern Russian city, and today covers a vast terrain of 640 acres. The site was originally managed by a pair of Dutch flower enthusiasts, the Ramm brothers. During the early years of the Garden, the Ramm brothers imported seeds and plants from France and Germany, growing exclusive flowers such as roses and hybrids, and cultivating a complex winter orangery. They handed management over to Rostov State University in 1928, but their legacy of biodiversity and ornamental flower cultivation has been maintained up until today.

Over its history, the Garden has faced a number of setbacks—several hundred of the seedlings and plants died in the planting transfer in the early days of the Garden. Ten years later, over half the Garden's collection was lost in the Second World War, and a long period of renovation and restoration work ensued. During this period, the Garden dedicated itself to the analysis and cataloguing of the amassed species, developing its collection and reworking the landscape and exhibits. Auxiliary industries were also established within the grounds to support its needs, such as a Pottery Workshop to produce growing pots, a metal workshop, and of course several greenhouses and nurseries.

The Garden now boasts over 5,000 types of tree, shrub and herbaceous plant, and over 1,600 species of flora from all over the world. A large portion of the Garden is given over to forestry, cultivating native forest types as well as creating artificial forests typical of other climates and countries. An area of over 24 acres has been cultivated to recreate the Russian steppe, in order to study and preserve the native flora and fauna indigenous to the habitat. The Garden also includes a collection of medicinal and aromatic plants, a floral collection, and a large collection of tropical and sub-tropical plants.

Great emphasis has been placed on conservation and research into biodiversity. However, as well as a scientific mecca, the Garden also provides a religious focus, thanks to a much revered monument built alongside a natural spring on its grounds, which was dedicated to Saint Seraphim of Sarov and has been consecrated by the Russian Orthodox Church.

tag and opposite bottom right: A promenade lined by Juniper Trees (*Juniperus virginiana*). photograph: AN Shmaraeva. opposite left: A variety of tropical succulent plants originating from Africa and South America, part of the Tropical Collection. photograph: TA Petushkova. opposite top: A Maidenhair Tree (*Ginkgo biloba*), situated in the Dendrological Collection. photograph: MV Kuropyatnikov. below: The Decorative Flowers Section of the Alpine (or Rock) Garden. photograph: AN Shmaraeva.

SINGAPORE
SINGAPORE BOTANIC GARDENS

Singapore's Botanic Gardens were established by the country's founder, Sir Stamford Raffles, in the early nineteenth century. Raffles was a keen amateur botanist interested in all aspects of plant evolution, whose book, *The History of Java*, earned him a knighthood. *Rafflesia arnoldia*, the plant bearing the largest flowers in the world, is named after him. Using an English model for his founding design, Raffles selected 47 acres on one of the small rises near Singapore Harbour for a garden that would explore the cultivation of economic crops such as cocoa and nutmeg. The site was planted with 125 trees, 1,000 seeds of nutmeg and 450 clove plants. With this initial foundation in place, the Garden became the supplier of plants to Singapore's first spice plantations. After Raffles death, however, the Gardens fell into disrepair, and was closed in 1829.

The Gardens at their present site was founded in 1859 by an AgriHorticultural Society. Planned as a leisure garden and ornamental park, the Gardens were very much inspired by European

models, with a carriage drive and a large selection of roses. In 1874, the Society handed over management and maintenance of the site to the colonial government, who deployed Kew-trained botanists and horticulturists to administer the Gardens.

The Gardens' first Director, Henry Nicholas Ridley, came to the Gardens in 1888 and brought it up to the highest of standards. He persuaded Malaya's planters to grow rubber trees, and through extensive research discovered a way to harvest commercial quantities of latex without harming the trees. He turned the Gardens' forest clearings and wasteland over to rubber, and when the automobile industry had a sudden rush for rubber, the Gardens had a ready source of seed supply. As a result, the Gardens' revenue increased dramatically, and the rubber plants became the basis for Southeast Asia's rubber industry, which sustained the region for years to come.

It was also during Ridley's time that the hybrid orchid, Vanda Miss Joaquin, was discovered, subsequently becoming Singapore's national plant. This entailed a lengthy history of orchid cultivation. Today, a hilly 7.4 acres of the Gardens have been allocated to a vast collection of more than 1,000 species and 2,000 hybrids of orchids. Within the orchid garden itself, there are a number of attractions such as the Orchidarium, the Tan Hoon Siang Misthouse (a colourful labyrinth of different hybrids), and the Coolhouse.

The Singapore Botanic Gardens are currently investing in botanical research, education and new public facilities. New attractions, such as the Ginger Garden, Evolution Garden and the Children's Garden are being added to keep the Gardens relevant as a leading destination. The Gardens are also home to a small tropical rainforest, of around 15 acres in size, which is older than the Gardens themselves.

tag: The Vanda Miss Joaquim, Singapore's national flower. opposite: A selection of Vanda Miss Joaquim Orchids growing in landscaped beds. above right: The Gardens' rocky lagoon. above left: View of the Evolution Garden.

SLOVAKIA
ARBORETUM BOROVÁ HORA

The Arboretum Borová Hora was established in the spring of 1965, when planting began on its hilly territory. It has since expanded to cover a site of 116 acres, focusing largely on the preservation and study of dendrological and forestry composition native to Slovakia. Searching for significant populations and rare forms of tree species in Slovakia is the arboretum's most important task and such species are propagated generatively and vegetatively, contributing towards a valuable archiving of the regional dendrological gene pool.

However, studying and maintaining local vegetation characters is not the sole function of the Arboretum Borová Hora. Associated as the botanical research and development workplace of the Technical University in Zvolen, Slovakia, its collections serve pedagogical, scientific and research purposes as well as providing a location for various cultural activities, and displaying various ornamental and international species of flora.

Situated three kilometers north of the centre of Zvolen, the Arboretum spreads out on the southwestern spurs of the Zvolen uplands. The geo-morphological character of this territory gives the Arboretum a varied terrain, with several areas of gentle and steep inclination. This, together with the altitude upwards of 300 metres, gives the landscape an upland, mountainous character, and it consequently enjoys geological and meteorological conditions suitable for the cultivated native forests.

The collection of this Arboretum can be neatly divided into three, comprising of the collection of tree species, that of roses and a third of cacti and succulents. The former is obviously the primary concern of the Arboretum and presently constitutes approximately 1,500 taxa; with regard to significantly deteriorating conditions of the environment, the collection of tree species in the Arboretum is receiving greater importance. Particular attention is paid to endangered taxa, within all the genetic permutations of various indigenous species. Much effort is spent on preserving and documenting the largest possible range of morphological deviations, as well as the examples of the original species.

The study of variation means that this Arboretum contributes invaluable and interesting data to the study of this phenomenon.

Though roses are not the primary focus of the Arboretum, they hold a great importance here too, and indeed the cultivation of rose species began at the same time as the dendrological plantings. Once again, particular attention is paid to the wide variations of species indigenous to the landscape of Slovakia. This provides an excellent and increasingly important base for future breeding, as many of the species cultivated here are particularly old native

strands, such as the garden roses of Rudolf Geschwind, that are not only rare, but also resist the unfavourable conditions very well. Study and genetic preservation of these strains can help create new varieties which can be used ornamentally in a variety of location and thus, their properties give them an arguably greater importance than recent cultivars. However, both old and new cultivars are enjoyed equally here in their attractive and fragrant plantations; over 300 of these are arranged in beds, whilst the garden rose collection stands at 150, further embellished by 60 trailing varieties and 90 more of miniature, but still perfect examples of these much loved flowers.

Continuing the theme of variation within species, the cacti and succulent collection is arranged to display the diversity within these groups of plants. A more recent addition to the Arboretum, at first it could present only a general character of this botanical category, housing just 100 genera. Now, however, it also specialises in North American and Mexican cacti, encompassing mainly the genera *Mammillaria* and *Echinocereus*, and currently displays 600 varied examples within the hundred types.

One of the principal aims of the Arboretum is the educational activity aimed mainly at the forest dendrology, therefore students of the Faculty of Forestry and the Faculty of Ecology and Environmental Sciences of the Technical University make use of the collections and the total area for training and experimental work. It contributes significantly to gene pool rescue and preservation of the indigenous dendroflora.

The technical staff provides professional consulting activities to the public concerning the greenery growing in gardens and houses, and the park naturally provides a very pleasant and informative ambiance in which to familiarise with native Slovakian vegetation.

tag: The Firecracker Plant (*Echeveria setosa*), which originates from Mexico. opposite top: Various varieties of Rhododendron. opposite bottom: The Rock Garden. photograph: Vladimir Jezovic. top: One of the Aboretum's Lilac (*Syringa vulgaris*) plants. photograph: Vladimir Jezovic. bottom: View of the Aboretum's Pavilion. photograph: Vladimir Jezovic.

LJUBLJANA UNIVERSITY BOTANIC GARDEN

The historic Ljubljana University Botanic Garden has had a chequered life story reaching back almost 200 years. Founded in 1810, at which time Slovenia was part of the French-controlled Illyrian Provinces, it was designated as a garden of native flora and a section of the École Centrale. The Garden grounds originally covered 33 acres, planned and directed by Franc Hladnik. Thanks to Hladnik and his connections with the Austrian botanists, the Garden was not closed down after the restitution of Austrian sovereignty.

After 1822 it was enlarged by half of its original size again and surrounded with a wall, until a further enlargement in 1834. One of Ljubljana's subsequent director of note was Andrej Fleischman, a pupil of Hladnik, whose discovery of Fleischmann's Parsnip (*Pastinaca sativa var.fleischmannii*) is one of the most interesting and indeed unique plants in the Garden. It was found on the grounds of the Ljubljana Castle in the first half of the nineteenth century but has since survived only in the Ljubljana Botanical

Garden. After a short lull in development, a Renaissance of the Garden began at the end of the nineteenth century, and the next few years saw the successful publications of annual seed indices and herbariums. Consequently, the Garden began exchanges with similar institutions internationally.

The University of Ljubljana was founded in 1919, and since 1920, has assumed management of the Garden. In the final stages of its evolution, after the Second World War, the Garden was once more enlarged and acquired a new greenhouse. Today, the Botanic Garden is a unit of the Department of Biology of Biotechnical Faculty, Ljubljana. Its plant collection includes more than 4,500 species and subspecies, a third of them being native to Slovenia and two thirds coming from various parts of Europe and other continents.

The arboretum occupies the oldest part of the original grounds, displaying many examples of trees and shrubs constituting the dendrological species native to Slovenia. Special habitats have been carefully created for the display of ecological groups and eco-geographic varieties, including marsh and water features for aquatic plants and a rock garden for plants originating from mountainous regions, allowing these to be easily viewed in naturalistic conditions away from their native locations.

Other sections of the Garden are variously arranged. The Plant System Garden presents plant families in the form of an evolutionary stem. The enormous utility of the plant world is highlighted in the Thematic Garden, where species are presented according to their properties, such as their therapeutic, culinary and industrial uses. The Greenhouse accommodates tropical, sub-tropical and Mediterranean plant species, which require higher humidity and a constant temperature throughout the year. For the Garden's practical purposes, the Cultivation Section houses plant beds for the cultivation and propagation of plants within the Garden itself. It is also reserved for scientific research, though public viewing is possible under the guidance of a member of staff.

The Garden also engages in scientific research and pedagogical activities, exchanging its *Index Seminum* with 293 botanic gardens around the world and playing a major role in the growth and protection of endemic and endangered species of Slovenia. A considerable number of these species have been successfully propagated and eventually re-introduced into their natural habitat.

tag: A Snowdrop (*Galanthus nivalis*). opposite: View of the Garden's Blooming Tamarisk (*Tamarix gallica*), Common Laburnum (*Laburnum anagyroides*) and Judas Tree (*Cercis siliquastrum*) in bloom. below: A section of the Rock Garden. all photographs: Jože Bavcon.

SOUTH AFRICA
KIRSTENBOSCH NATIONAL BOTANICAL GARDEN, CAPE TOWN

Showcasing only indigenous South African plants, the Kirstenbosch National Botanical Garden is probably the most vernacular of all the world's botanical institutions. Despite being an important junction for European ships circumnavigating towards Asia up until the opening of the Suez Canal in 1869, the Garden in Cape Town remained virtually untouched by the stashes of foreign flora taken home by Europeans from the Far East. Bearing in mind how indebted gardens such as Kew, Berlin, and Oxford are to the colonisers' efforts, it is interesting to note how Kirstenbosch has long remained steadfastly native.

This might have been something to do with the time-frames in which the Garden developed. Although the site of Kirstenbosch, on the eastern slopes of Table Mountain, has been used for the inadvertent cultivation of plants since 1660 (when a hedge of wild almond and brambles was planted to form the boundary to protect the colony's cattle, sections of which still exist today), it was not until 1913 that the Botanical Garden officially underwent development. Despite previous dominance by the Portuguese, the Dutch and British forces, by 1913, the increasingly independent stance of the South African population must have had an effect of the patriotic curatorial stance at Kirstenbosch. Furthermore, with ten per cent of all known plant species on earth growing naturally in South Africa, there was probably little need to look abroad.

It was Professor Pearson—Chair of Botany at the South African College—who took control of the initial design and establishment of the Garden. There was no budgetary room for a salary, so he worked in an honorary capacity, starting first with what is known as The Dell. This section had already been developed in the early nineteenth century under the English Occupation, when Colonel Bird built a house on the site, planted chestnuts and built a bath in the Garden. By the time Professor Pearson took control of the area, the estate had changed hands twice and had slowly disintegrated. In 1913, when Pearson started work on the Garden, he faced numerous obstacles.

However, the mission Pearson set himself and his team slowly took hold of the wreck. It was Pearson who settled the garden's ethos to preserve South Africa's unique flora, and to cultivate only indigenous plants, and it was he who was responsible for the hundreds of wonderful savanna, fynbos, karoo, and cycads that are now housed at Kirstenbosch.

The curators here are still heartily keen on reminding visitors exactly where they are and what culture they are in. Architecturally speaking, the Garden is structured by the cobbling, curbing and dry stoning of local Table Mountain stone and, in terms of planting, most of the separate gardens establish an unbreakable bond with the native land. The Peninsula Garden displays some of the 2,500 plant species found indigenously on the Cape Peninsula, and the Water-Wise Garden demonstrates how to create a garden that needs far less water and maintenance than usually required for cultivation. In reference to the desert climate of South Africa, this is another feature that emphasises the fact that, above all, this garden is explicitly South African in its nature and in its spirit.

tag: Mandela's Gold (*Strelitzia reginae*), a yellow variant of the orange-flowered Bird-of-paradise (*Strelitzia reginae*). photograph: SANBI. opposite top right: The Botanical Society Conservatory, a desert house that displays the arid plants of South Africa, as well as bulbs, ferns and South African alpines. photograph: Ernst van Jaarsveld. opposite bottom left: Eighteenth century plan of the Dutch East India Company Garden. opposite bottom right: *Erica verticillata*, a decorative member of the Heath family, which is now extinct in the wild. above: The Garden in summer. photograph: Alice Notten.

The Witwatersrand National Botanic Garden, located just outside Johannesburg was recently renamed the Walter Sisulu National Botanic Garden, in honour of the freedom fighter who spent over 20 years in prison during the apartheid era. The Garden sits on a massive 741 acres of land—75 of which is landscaped, and the remaining 666 acres is left wild. At the heart of the Garden is the magnificent Witpoortjie Waterfall, which tumbles straight down a vertical cliff into a lake below. The Walter Sisulu Garden is one of eight botanic gardens managed by the South African Biodiversity Institute (SANBI), and was founded in 1982, although the area was always one of outstanding natural beauty, and so had been frequented by visitors for many decades before.

The uncultivated areas of the Garden are a mosaic of grassland and savanna, collectively known as "Rocky Highveld Grassland". The dense bush accommodates over 600 native plant species including Bushman's Poison (*Acokanthera oppositifolia*), River Bush-Willow (*Combretum erythrophyllum*) and Coral Tree (*Erythrina lysistemon*). It is also home to an abundance of wildlife including jackals, antelope, and a breeding pair of Verraux's Eagles in the cliffs by the waterfall. Over 230 species of birds have been recorded in the Garden.

Carefully designed walks through the grounds allow visitors to partake in the incredible beauty of the area—guiding them through the landscaped section to the nature reserve, where they can witness both the enormous botanical scope as well as the fascinating geology that the Garden incorporates. An environmental education centre, a succulent garden, a cycad garden, a fern trail and an arboretum are just some of the points of interest that the Garden has to offer, and plans to develop an information area and further visitor facilities are currently in the pipeline.

tag: Cooper's Grass Aloe (*Aloe cooperi*). opposite top: The Peoples' Garden surrounded by Krantz Aloe (*Aloe arborescens*). opposite bottom left: The Clivia Walk, which leads into a natural forest along the Crocodile River. opposite bottom right: *Ensete ventricosum,* a wild relative of the Banana Plant. all photographs: SANBI.

SOUTH KOREA
SEOUL NATIONAL UNIVERSITY ARBORETUM

Belonging to Seoul National University, the Arboretum was established in 1967. It is divided into two sections; the Kwanak Arboretum and the Suwon Arboretum.

The Kwanak section of the Arboretum is by far the larger, reaching from the southern suburbs of the South Korean capital to the Anyang Valley of Mount Kwanek, covering over 3,707 acres. It is surrounded by a natural oak and maple forest, and includes over 1,700 species of tree and plant life. While the Arboretum endeavours to sample flora from all over the globe, South Korea's naturally cool, temperate conditions favour hardy shrubs and trees and so notable success and interest is derived from the plantation of species from similar climates in the northern hemisphere.

The Arboretum forms part of the Agricultural and Life Sciences campus of the University. At 55.6 acres, it contains 500 species of plant life and provides a vital resource for those studying at the college. Approximately 25 acres of this section have been cultivated into a pine forest ecology. The Arboretum also features an aquatic botanic garden, added in 1985, for the study of water-based plants and plant life native to wetlands habitat. Several greenhouses allow for the cultivation of seedlings and subtropical and tropical plants, which would not otherwise survive the cool weather system endemic to South Korea.

The gardens are intersected with immaculately tended paths bordered by ornamental planting, so that visitors can enjoy the scenic terrain and the vistas of the mountainous countryside. Following on from the native and Japanese influenced traditions of botany, landscaping and gardening, the gardens are devoted to forestry, research and education initiatives, as well as providing environments that can be enjoyed by its visitors. The mission of the Arboretum is to engage in the acquisition, display and conservation and to study collections of botanical examples, sharing the information with national and worldwide institutions.

tag: The Korean Abelialeaf (*Abellophyllum distichum nakai*), an endangered species native to Korea. opposite: The Arboretum Greenhouse. above: A pathway leading through the Perennial Garden. below: The Grass Park, a large open lawn area within the Arboretum.

BOTANIC GARDEN OF MADRID

The Botanic Garden of Madrid was founded in 1755, near the Manzaneres River. In 1774, however, King Carlos III relocated the Garden to its present site, the Paseo del Prado, in an attempt to centralise scientific institutions. To this day, the Natural History Museum, the Astronomic Observatory and the Botanic Garden all share a site. The design of the Garden was originally given to the royal architect, Francesco Sabatini, who divided the terrain into three terraces, but in 1780, the project was handed over to another architect, Juan de Villanueva, who gave the Garden a more explicit Neoclassic temper. He was also responsible for the Garden's centrepiece—the beautiful Villaneva Pavilion, declared a historical monument in 1942.

The history of the Garden very much reflects the history of Spain. The last two decades of the nineteenth century and the first years of the twentieth witnessed promising activity in every respect in which a botanic garden is concerned: botanical research, education and horticulture—an amazing 3,000 species were cultivated during the Garden's infant years. The botanical studies benefited greatly from overseas expeditions that generated priceless historical collections of herbarium specimens and plates. This period culminated with the directorship of the well-known botanist Antonio Cavanilles, who studied and published books on of the botanical specimens brought from the Americas, including the horticulturally important genus Dahlia. During the first half of the nineteenth century, the Garden languished, a reflection on the unstable conditions in Spain following the invasion by Napoleon and the disputes between liberals and absolutists.

The second half of the nineteenth century improved the state of the Garden in some important ways: a new greenhouse was constructed and the upper terrace was redesigned into a new Romantic style. This period of prosperity carried through until

the 1920s, but was brought to a crashing halt in 1936, with the outbreak of the Spanish Civil War. The international isolation that followed reduced the resources for research and horticulture, and the grounds deteriorated to the point that the Garden was closed to the public in 1974.

Towards the end of the 1970s, the political authorities relented, and granted the financial support required to rejuvenate the Garden. Architect Leandro Silva implemented the plans, preserving the two lower terraces in the Neo classical spirit they had been built in, and renovating the upper terrace in a nineteenth century romantic style. The Villanueva Pavilion was also restored, and the Garden was reopened in 1981.

The Garden today is much celebrated for its unique geometric design, which synthesises two and a half centuries of history. In recent years the garden has gone from strength to strength. As well as research and educational achievements, the Garden has seen a new exhibition greenhouse built (1993) and a slope

reconfigured into a new space for visitors (2005). Due to its central location in the cultural heart of the city (in close proximity to the main museums—Prado, Reina Sofia and Thyssen), the Garden has almost half a million visitors a year, and is a small but extremely popular cultural institution.

Especially worth seeing are some of the trees that have withstood all the highs and lows of the Garden's history, including a 40 metre tall *Zelkova carpinifolia*, a 240 year old *Cupressus sempervirens* and an *Ulmus minor* whose trunk is seven metres in diameter.

tag: Bromeliads, orchids and other epiphytes in the Tropical Humid Section of the Greenhouse. photograph: Esther García Guillén. opposite: A view of the small pond by the Laurel Terrace, where the Bonsai Collection is exhibited. photograph: Esther García Guillén. above left: The Horse Chesnut Grove, located in the upper terrace. The fountain in the foreground was discovered during the Garden's restoration in the 1970s. above right: A Vine Arbour, constructed in 1786. photograph: Esther García Guillén. middle: A view of the upper terrace the Garden's Greenhouse, built in 1856. photograph: Esther García Guillén. below: A display of tulip cultivars arranged in rows on the Lower Terrace. Photograph: María Bellet Serrano.

A charming variety of cobbled trails and tropical succulents, the Viera y Clvijo Botanic Garden offers a generous selection of island plant life. Located in Tafira Alta near the capital city of Las Palmas, the Garden covers 66.7 acres is the largest botanical garden in Spain.

The Garden has a unique collection of native Canarian flora, which has mostly disappeared from mainland Spain where it originated. The Garden was established in 1952 by the Swedish botanist Eric Sventenius, who studied at various universities across Europe, including Spain where he studied at the Marimurta Botanical Garden in Blanes. In 1931 he traveled to the Canary Islands and in 1952 he began working at a botanical garden in Tenerife, the Jardín de Aclimatación de la Orotava, where he studied and catalogued unclassified Canarian species. Wanting to create a botanical garden strictly dedicated to Canarian flora, he founded Jardin Canario in 1952 and the doors opened to the public in 1959. The Garden was named after the botanist Jose de Viera y Clavijo, the professor best known for his exhaustive work, *History of the Canary Islands*.

The Garden is dedicated to all the plants indigenous to the seven islands of the archipelago and the islands belonging to the Macaronesia Islands, which are Madeira, the Azores and Cape Verde. At the centre of the Garden is a bust of Viera y Clavijo standing proudly above the grounds in a centre plaza surrounded by native palm trees and succulent species grouped according to their island of origin. Plants live in their natural environment, growing between rolling rocks and leaning hills. Small round cacti bloom on the grounds of the cacti and succulent garden where over 2,000 verities of the plants live. A curving wooden bridge guides visitors to a magnificent garden of ornamental plants, featuring boldly coloured local plants, located around the Plaza de Fernando Navarro. Despite the naturally tropical climate, a large greenhouse produces a warmer and moister climate showcasing a variety of tropical succulents. The greenhouse is named for Zoe Bramwell a prominent botanist of the area who co-wrote *Wildflowers of the Canary Islands* among other books.

A stone bridge designed after a bridge that once crossed the Guiniguada Ravine in Las Palmas takes visitors through a meandering tour of local plants.

Another homage to naturalists of the Canaries, The Fountain of the Wise, represents ten botanists who dedicated their work to the flora of the Canary Islands. Yet the winding gardens are not simply a display of lavish tropical flora, there is also a native Canarian pine forests within the garden. The nearby Laurisilva forest showcases the bay laurel tree, which covered the island during pre-Hispanic times, but has since been mostly eliminated from the local landscape. There are also dragon trees, shrubs and herbs living in the forest. Canary Palms, including one large sample of the species sit on the Matias Vega Square.

Music and dancers can be heard in the Plaza de los Nenufares (The Water lilies Square) when the garden hosts a number of concerts, folk dances and ceremonies throughout the year. The farm tool shed is a traditional Canarian building restored in 1989, which displays historical farm and garden tools.

Research at the Garden is focused on determining the origin of Canarian species and their relationship to other flora. Because there are a number of endangered plants in the Canary Islands, researchers work to understand the classification, biodiversity and reproduction cycles of native plants. The garden also has a Seed Bank which started in 1983 and aims to create a bank of Canarian and Macaronesian region plants. The bank currently holds more than 2,000 samples of wild seeds from about 400 endemic species. As some of the plants are highly threatened, the Seed Bank provides storage for a wider variety of genetic variability than that of the local plant population.

tag: A spray of Flor de Mayo or Mayflowers. Originally a wild shrub, the flower is now cultivated for domestic ornamental use. opposite top left: Various types of succulents and cacti thriving in the Islands' sunny climate. photograph: Angel Luis Alday. opposite top right: A waterfall within the Botanic Garden. photograph: Manuel Quevedo. opposite bottom: View of the lawn leading up to a banked plantation of native palms and ferns. photograph: Juan Manuel Lopez.

Franz Bauer Ferdinand Bauer Norbert Joseph von Jacquin Jan Gronovius Emperor Franz John Sibthorpe Thaddäus Botany Bay Ellis Rowan Franz Bauer Ferdinand Bauer Norbert Boccius

Where Art and
Science Meet

Where Art and Science Meet

Margaret Stevens

When approached to write something on the link between botanical artists and botanical gardens it was difficult to know where to begin. Not because of a dearth of material but rather the opposite, as over the centuries almost every botanical artist of note has had some connection with the great botanical gardens of the world. Many of these men and women are very well known but I have tried to introduce a few less familiar names because of the way their lives are interwoven.

So I will spin a web, starting one October day in 2000 when I was fortunate enough to be in the archives at the Real Jardín Botánico in Madrid. I had looked at a number of beautiful pieces of botanical illustration when something very different was put before me. This was a hand-painted shade card with 120 segments of colour, each one numbered. I was told it had belonged to one of the Bauer brothers—which one, Franz or Ferdinand, was immaterial as I was awestruck. So this was how they managed to sort out their colours for the massive body of work which both of them produced throughout their lives. It was obvious to any painter, even a lesser mortal like me, that they had not enjoyed the luxury of sitting before their specimens for days on end, completing each study from initial sketch to finished picture. In any case, flowers fade fairly quickly and the soft sound of falling petals, long before work is completed, is all too familiar. These days most of us mix and match colour at the start of the painting and rely on that to see us through. The Bauers did not need to do that but simply matched each part of the subject to their shade cards and numbered the drawing accordingly. They could then safely return to work on it as time and circumstances allowed. Seeing this practical example of their art meant more to me than any amount of fine paintings that morning, and I could not help but wonder how it had arrived in Madrid, when neither Bauer had any connection with the Spanish capital.

Portrait of Nikolaus Joseph von Jacquin, 1727–1817, explorer and accomplished botanical artist.

My search for answers took me back many years and introduced some of the people whose influence shaped various careers, particularly that of Franz Bauer, 1758–1840. He was one of six surviving children whose father Lukas was Court Painter to the Prince of Liechtenstein. Lukas died when Franz was only four years old, leaving his widow to bring up a very young family. She obviously did a good job of this, not least by encouraging them all to draw and copy artwork, probably putting their father's materials and papers to good use. Her four sons had surely inherited his talent, a point worth making when so little value seems to be placed on natural talent these days. In the case of the Bauers, both nature and nurture played their part.

They were also fortunate to be growing up in Feldsberg, close to the Czech border, where there was a monastery of the Brothers of Mercy, the hospital Order of St John of God. Although one cannot be sure, it is reasonable to assume that the

Bauers received some kind of education there although it might not have been of conventional kind, as here Franz and his brothers met the second powerful influence in their young lives.

Enter Norbert Boccius, a member of the Order, who had qualified as a doctor of medicine in 1763 at the age of 22 and moved from Vienna to the monastery at Feldsberg as professor of anatomy, surgery and nursing. He was never ordained and in the tradition of the Order, Feldsberg housed only one priest, the remainder being lay brothers, Boccius amongst them. There was, however, a hospital and nursing school which was remarkably ahead of its time by specialising in the training of male nurses. In monastic tradition there was a garden in which to grow medicinal herbs for the treatment of patients and the nursing school. This was ideal for Boccius who was already renowned as a botanist and in the absence of facts it is safe to assume that young Franz was encouraged by Boccius to the study of botany allied with art. It is also very likely that during this period they attained the capacity for hard work which persisted throughout their lives, as can be seen in the vast amount of work produced. Certainly, if they had been encouraged to help out in the hospital, it would have introduced them to the realities of life and death.

Today it is often said that success in any field is dependant on whom one knows. This was even more true then than it is now. Certainly Boccius was a well known

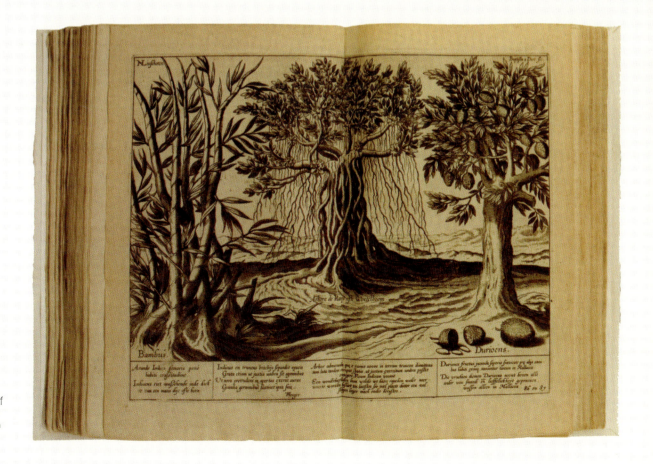

An early example of botanic illustration depicting European Travellers.

Early work by acclaimed botanical artist Francis Bauer, court painter to the Prince of Liechtenstein.

figure and the circle of gardeners, artists and botanists radiating from Vienna would draw the talented Bauer brothers in. By 1770 he had already held the position of sub-Prior once and Prior twice, now in what might be seen as a 'rest' period he began on his great work, the *Codex Liechtenstein*. This 14 volume florilegium was to take over 30 years to complete, the final volume being handed over to the Prince of Liechtenstein in 1805, only a few months before Boccius' death in December of that year. What seems so remarkable in our day is that teenagers like the Bauers were mature enough to assist with such important work. This could provide much food for thought and discussion, but it is time to introduce the third influential figure.

Nickolaus Joseph von Jacquin was born in Leiden in 1727 and attended the Jesuit school in Antwerp until changes in the family fortunes meant that he had to look on education as a stepping stone towards a career, rather than something which could be indulged for its own sake. In January 1745 he enrolled at the University of Leuven where he became great friends with the son of Linnaeus' patron, Frederik Jan Gronovius. These teenage boys enjoyed many botanical excursions culminating in a visit to the Leiden University Botanical Garden, where the sight of one particular flower, *Costus speciosus*, so enraptured young Jacquin that at that moment a love of botany, and in particular botanical art, was born.

I quote here from his memorial address given in Vienna in 1818: "… he found himself… gripped by admiration for Nature… and for Science, which was capable of capturing its essence in mere words, and of immortalising its shape and colour by means of art".[1] How accurate this is, since it refers to something which happened over 70 years before, there is no way of knowing. What is worthy of mention is that this is one of the early occasions when the words art and science were used together in an attempt to explain botanical art. It remains the best description available to this day.

Jacquin pursued his botanical interests whilst still studying medicine in Paris but he was constantly hard-up. Fate took a hand when Gerard van Swieten, the former family doctor and long-time friend, came to his rescue. Van Swieten had moved to Vienna on his appointment as physician to Maria Theresia and now invited the 25 year old Jacquin to join him. This would allow him to continue his studies without financial worries and eventually to qualify. Wisely Jacquin accepted the offer but was soon disappointed in the standard of tuition available, nor were there gardens equal to those in Paris, in which to nurture his thirst for botany.

This would change when, in 1754, the University acquired the first plot of land for a garden and in the following year Emperor Franz I acquired more land for a so-called Dutch Garden at Schonbrunn. This would be somewhere for Jacquin to study plants and classify them, as gradually rare specimens were introduced and a hothouse built which enabled tropical plants to be cultivated.

An early work by famed botanist Carolus Clusius depicting *Nymphaea*, which he included in one of early botany's most important works, *Rariorum plantarum historia*.

Not surprisingly, with so much time being taken up with botanical studies, Jacquin had still not qualified when he was asked to travel to America on a plant hunting expedition for the Imperial collection. One can imagine his delight and he immediately enrolled at the Imperial Royal Drawing Academy in order to improve his artistic skills in readiness for the trip.

The expedition lasted for five years, during which time the party suffered illness, piracy and the horrors of travelling on a slave ship, but was nevertheless extremely successful. Apart from many plants which helped to make Schonbrunn second to none, Jacquin also collected live animals and mineral specimens. There is no doubt that his fame increased as a result of this trip.

On his return, Jacquin turned his attention to illustrating a book based on the American expedition, which would contain over 180 large copperplate illustrations. This industrious man never faltered, continuing to help stock the Botanic Garden with seeds and plants through his growing number of connections in Europe, until in 1763 another change of direction took him to Schemnitz and an appointment as professor for practical mining and chemistry. Somehow he found time to marry and have three children, the eldest of which, Joseph Franz, born in 1766, would eventually succeed him as director of the University Botanical Garden.

Hand-painted shade card belonging to one of the Bauer brothers. The 120 different hues were used to mark botanical paintings so the work would accurately reflect the living flora. Courtesy of Archivo del Real Jardín Botánico, Madrid.

In 1769 he returned to Vienna, this time as professor of chemistry and botany and, of greatest importance to this story, as director of the University Botanic Garden and advisor to the Imperial Garden at Schonbrunn. At Schonbrunn he turned his hand to everything from sowing seeds to supervising the publication of numerous books on botany and medicine. Above all, based on his own artistic abilities, he supervised the standard of illustrations. And, he finally qualified as a doctor at the age of 41.

It is easy to see how Jacquin and Boccius would have had much in common and the Bauer brothers, making field trips in search of plant material from their base at Feldsberg became part of the circle, working on the *Codex Liechtenstein* and eventually coming under the patronage of Jacquin. It would be easy to lose sight of how young the brothers were at this time and during the 1770s it is likely that they were still being guided by Boccius because not until 1781 does the name of Franz Bauer appear in the register of the Vienna Academy of Fine Art. That same year elder brother Joseph left the Academy to continue his studies in Rome. In that year the brothers left home and Ferdinand lived with Jacquin in his official residence. One imagines Franz either lived there too or else was a frequent visitor. Apart from the illustration work which they undertook for Jacquin, Franz also prepared demonstration material for use at lectures. It is worth mentioning here that by 1784 around 1,500 paintings had been completed towards the 2,748 in the finished *Codex*. The majority of the 1,500 were painted by the Bauer brothers. After Joseph's departure for Rome, Franz and Ferdinand had carried on with the work together until Ferdinand left in 1786 to join John Sibthorpe, professor of botany and director of Oxford Physic Garden, as illustrator on his trip through the Ottoman Empire. For this Ferdinand was paid 80 pounds per annum.

For another two years Franz carried on working on the *Codex* until his opportunity came to travel with Jacquin's son, Joseph Franz, as friend and companion, on a grand tour of Europe with the emphasis on visits to botanic gardens for study and to find interesting plants which could be sent back to the head gardener at Schonbrunn. When they left, Franz could not have known that he would never return to his homeland or see his mother or Jacquin again.

Franz eventually reached England and settled at Kew in 1790. Thanks to the patronage of Sir Joseph Banks he was able to work as resident draughtsman at the Royal Botanic Gardens for 40 years. He enjoyed the title of botanical painter to George III whose wife, Queen Charlotte, benefited from his instruction in the art of botanical illustration as did William Hooker, a future director at the Royal Botanic Gardens.

There was plenty of work to keep the industrious Franz happy in the forthcoming years with new plants arriving from around the world. He was almost certainly the first person to record accurate dissections, which were soon recognised as an important aid to the study and identification of species. Even a grain of pollen was not too small for his attention and he would illustrate these at a high magnification. It is unlikely that microscopes were sufficiently developed for this purpose and it is quite possible that he made use of the camera lucida which

Work depicting Grevillea banksii made by Ferdinand Bauer during the voyage of *Flinders* between 1801 and 1803.

was a recent invention. This shows a readiness on Bauer's part to embrace new technology, in the same way we now believe that great artists such as Vermeer adopted the camera obscura.

As Bauer worked at Kew, one can imagine his thoughts turning to the *Codex* and the artists who were working towards its completion. Sadly this magnum opus was never put on public display and remained in storage in Vienna from 1805 until 1944 when it was moved to Vaduz for safety. After the war it was transferred to Schloss Vaduz, home of the Princes of Liechtenstein. What did Norbert Boccius receive in return for this magnificent gift? For the first 12 volumes, two hospital beds were endowed by the reigning Prince for the monastery at Feldsberg, for as long as the

The legendary Marianne North pictured painting from live specimens, image courtesy of the Royal Botanic Gardens, Kew.

monastery existed. For the final two volumes—nothing. Boccius can truly be said to have undertaken and paid artists for this work out of his love for the subject. Artists then, as now, loved their work and frequently undertook it for a pittance.

Bauer would eventually give his collection of paintings to King George IV and when they passed to Queen Victoria she presented them to the British Museum. From there they were transferred to the Natural History Museum in 1881. Francis Bauer— for he had anglicised his name—was by all accounts a popular, indeed, loved man who when he died at the age of 82 was honoured by a handsome memorial in St Ann's Church at Kew, where he was interred in the churchyard.

So what about the shade card? One of Jacquin's pupils was Thaddäus Haenke, 1761–1816, who, according to his age, was a contemporary of the Bauers. This shade card accompanied him on his world tour, during which he travelled through the Magellan Straits into the Pacific and on to Botany Bay. Eventually, Haenke reached Peru where he went ashore—and died. Some time later his effects, for reasons unknown, were returned to Spain and the recipients, on settling his estate, donated the colour code to the Real Jardín Botánico in Madrid where it entranced me nearly 180 years later. The colours have been checked against known Bauer work and match perfectly.

Later in the nineteenth century it was a woman whose name would become associated with Kew, even by many who know little about botanical art, the redoubtable Marianne North. Whether you love or hate her work, and it is not difficult to find those who will speak from either viewpoint, the fact is that she was a truly remarkable woman, highly industrious with a keen enquiring mind, which I am sure would have endeared her to Franz Bauer.

Portrait of Marianne North, 1830–1890.

Born in Hastings in 1830 of a well-connected family, which was to provide the launch pad for her later travels, she enjoyed all the privileges of a well-brought up young lady from that era. Moving between town and country houses, life must have been idyllic. For a good deal of the time she had the company of an adored father, who had given up his parliamentary seat when in poor health. From the age of 16 to 19 she toured Europe, spending eight months in Heidelberg and undoubtedly this, and her enjoyment in reading at school in Norfolk, provided a better education than any governess could have done. Riding, singing and painting, with lessons from a Dutch lady filled her days. This came to an abrupt end when she was 25 following the death of her mother. By this time her father had re-entered Parliament and, having promised never to leave him, she moved with him to live in a flat in Victoria Street. Marianne was now able to obtain specimens to paint from Kew, where she made the acquaintance of Sir William Hooker.

Long parliamentary recesses were ideal for Marianne, her sister Catherine and their father to indulge their love of travel and summers spent in England were replaced by more adventurous trips to the continent, Hungary and Egypt. Marianne was 40 years old when, on a trip to the Alps, her father was taken ill. Luckily they managed to get home before he passed away. The loss of the man who had been her best friend and companion was a bitter blow and as soon as his affairs were settled

A particularly charming example of Marriane North's botanical paintings, Image courtesy of the Royal Botanic Gardens, Kew.

she went to Mentone on the French Riviera. With just a servant as companion Marianne threw herself into her work, painting from nature, in an attempt to "… try to learn from the lovely world which surrounded me there how to make that work henceforth the master of my life".[2]

This strategy worked and two months later they left Mentone and set off on a cruise around Sicily. The pattern was thereby established for the next 14 years, starting with a trip to the United States and Canada in 1871 in the company of a friend. Like all her journeys this one ranged from one extreme to the other, from the fear engendered by sailing through icebergs to another state of nervousness when invited to dine at the White House with President Grant. It later transpired that this had only happened because she was believed to be the daughter of Lord North, the former Prime Minister. Had this been so, she would have been a very old lady indeed! Her tireless sightseeing took her to Massachusetts, New Hampshire, New York and Niagara Falls, all the time painting plants and scenery along the route. She arrived back in England in June 1872 after spending some months in Jamaica where she found many exotic blooms to record.

Unlike what is sometimes looked on as a 'true' botanical illustrator, Miss North was also a landscape painter who managed to convey a sense of place to very many of her flower paintings. Working in oils she was able to quickly capture the habitat or even the entire surroundings as a background to her subject, not to mention a humming bird or some other variety associated with that particular vegetation. This is no doubt what upset her detractors, such as Wilfred Blunt, who wrote scathingly of her work in *The Art of Botanical Painting* and it saddens me to say that I suspect his attitude would have been different had he been writing about a man.

In the fashion of the day, Marianne North also kept a detailed journal, which was edited and published as *A Vision of Eden* by the Royal Botanic Gardens, Kew. The descriptions of her travels are a joy to read. However, her trip to Australia in 1881, which followed a suggestion made to her by Charles Darwin, gave me the clue I had been seeking and leads me on to the next thread in the web. Wherever she went, Marianne North unfailingly visited the botanical gardens, which in Australia included Brisbane, Sydney and Melbourne. The Brisbane Gardens did not impress her whereas those at Sydney she considered "lovely" and the ones at Melbourne even more beautiful. She recorded these in a small painting and the grounds at Brisbane provided the background for a scarlet passion flower. This did not prevent her from later writing to Sir Joseph Hooker at Kew, apologising for not painting all the flowers on his list, saying that "... the gardens did not tempt me—they are all so stiff and young".[3]

Portrait of Ellis Rowan in her wedding dress, 1873, portrait by J Botterill. Image courtesy of the National Library of Australia.

More to her taste was the time spent staying in a cottage at Albany, Western Australia as the guest of the lady she refers to as "Mrs R, the flower painter I had heard so much of".[4] There she was at last—Ellis Rowan, the artist who came to my attention a number of years ago when an Australian client sent me a book devoted to her work. When I read it I was immediately struck by the similarity of her life to that of Marianne North, not to mention her work.

Born in Melbourne in 1848 the eldest of seven children she was actually named Marian after her mother but used her second name, Ellis in honour of her Irish grandmother, the illegitimate daughter of King George IV (the recipient of Franz Bauer's paintings). Her education included the ladylike pursuits of embroidery, singing and painting in watercolour and was followed by a trip to England, when she was 21, where she stayed for a year. She was painting at that time and it is fair to assume at least one trip to Kew would have been on the itinerary. She won a prize for paintings of Australian wild flowers at the Melbourne Exhibition in 1872 and her interest in botanical art surged when her father, Charles Ryan, who was a keen botanist, bought 26 acres on Mount Macedon on which to build. The house, Derriweit Heights, became one of Australia's finest properties not least because of its splendid garden in the creation of which Ryan was advised by the eminent botanist Ferdinand von Mueller.

The Botanic Gardens at Brisbane were 'young' as Marianne North had observed. Created in 1846, von Mueller became the first director in 1857 and set about establishing a herbarium and scientific centre as well as building a remarkable plant collection. There was no one better to make suggestions and offer help in stocking the garden on Mount Macedon. It was also a boon for Ellis who was asked by von Mueller to paint specimens for his own collection. This is the sort of lucky happenstance which every botanical artist yearns for. The actual layout of the garden was designed by WR Guilfoyle who became the second director at the Botanic Gardens in 1873. With two such illustrious men involved in their development it is hardly surprising that Marianne North considered Melbourne to be the best of the three she had visited in Australia to date.

A typical example of
Ellis Rowan's botanical works,
unique for their scenic qualities.

Also in 1873 Ellis became engaged to a British Army Officer, Captain Frederick Rowan and following their marriage, settled in New Zealand where he became sub-inspector for the Constabulary of Armed Forces. This was a difficult time for Ellis, far from friends and family but with her husband's encouragement she devoted her time to painting the wild flowers which were all around her. It was 1878 before the couple, now with a small son, could return to Melbourne. Ellis accompanied her husband on his trips around Australia which gave her the opportunity to paint many indigenous flowers. It was on one such trip that she invited Marianne North to stay and the visit was obviously a great success. Both women admired each other's work and benefited from the other's experience. Ellis was able to show Marianne wonderful wild flowers to paint and in return it would appear she was encouraged to develop her own skills. Up to that time the younger woman had painted in gouache "… most exquisitely in a peculiar way of her own on gray (sic) paper".[5] It would appear Marianne North encouraged her to paint flowers in their natural habitat with vegetation and a landscape background. In this both women were following the work of Robert Thornton in his florilegium *The Temple of Flora* which showed exotic plants in magnificent settings rather than formal studies of plants with dissections on plain paper. The arguments as to the rights and wrongs of this continue to this day. My personal belief is that there is room for both expressions of botanical art so long as the plant is accurately depicted in every way. Common sense decrees that only an idiot would use a painting by Marianne North or Ellis Rowan to illustrate a treatise on, say, *Lilium* species, whereas in a work on plants from the rainforest they would be ideal since they conjure up a sense of place. Studio portraits of plants, however beautiful, will not convey the humidity of dripping foliage or the tiny, nearly extinct specimen nestling amongst mosses and ferns which could so easily be overlooked. It should also be remembered that, at the time these women were working, most people lived in a visually limited world. We take for granted the plethora of images which assail us in glorious colour but the Victorians had no bewhiskered version of David Attenborough to bring the jungle into the living room. I am sure that by the time Marianne North's paintings were hung in her gallery at Kew all those who could went to marvel at the things that she had seen and captured for posterity. But I jump ahead of myself so back to Albany!

In those days spent together the older woman also encouraged Ellis to travel and, it seems, to paint in oils, so she was certainly an influence in her career and development.

Some years before whilst on a visit to Japan Marianne had suffered a bout of rheumatic fever which had weakened her general health so in 1885 she returned to England for the last time. The Gallery at Kew had been opened in 1882 but soon she was too tired to visit it, that most generous gift which only the most unkind individuals describe as an egocentricity. Then, as now, success attracts sniping, usually borne of jealousy. Marianne retired to Alderley in Gloucestershire and took pleasure in making her garden as perfect as possible. Sadly, she soon became ill and died on 10 August 1890.

On the other side of the world Ellis Rowan spent the 1880s working incredibly hard, painting rarities for von Mueller, exhibiting and selling her work, both originals and commercially as published engravings. She made six trips to Queensland where

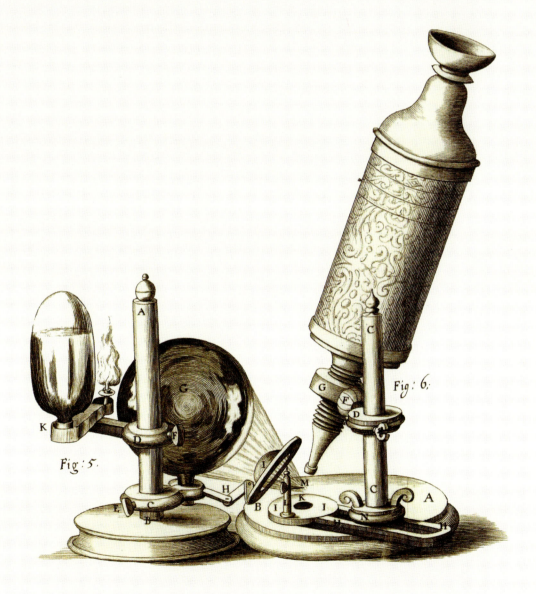

The compound microscope was instrumental in the advancement of botanic scientific drawings during the seventeenth century.

she was captivated by the plants of the jungle and produced what is considered some of her best work, behaving at odds with her delicate appearance and more like a female Crocodile Dundee, swinging on ropes over ravines, bitten and stung by insects and contracting fevers. She worked quickly in situ often under extreme difficulties of heat and humidity not to mention the ever-present threat of snakes and other reptiles.

In 1888 Ellis had a taste of male jealousy when she was awarded first place for a painting of chrysanthemums at the Melbourne Centennial International Exhibition against competition from all over the world. This was an oil painting and she had never exhibited in that medium before. She was totally unprepared for the outcry that this caused since her work contained flowers and birds so it could not be 'high botanic art'. She very sensibly rose above it and went on painting, becoming a household name as a result.

In 1895 Ellis came to England and within one year had held a very successful exhibition in New Bond Street. Queen Victoria was an admirer and kept three of her paintings to be made into a screen for her own room. Ever seeking pastures new, Ellis went on to the United States where she stayed for seven years. She travelled extensively and exhibited up to 500 paintings at several prestigious venues in Chicago and New York. Her attempts to find a home for her collection in America were unsuccessful and in 1905 she returned to Australia for the first time in ten years. Further expeditions ensued including two to New Guinea, chiefly to paint the birds of paradise, when she was approaching 70. Malaria eventually took its toll and kept her at home, where she campaigned for her work to become a National Collection. She was unsuccessful until 1923, when a year after her death the government agreed to purchase 947 watercolours for £5,000—a quarter of the asking price.

I would like to bring the story up to date by writing a few words about today's connection between artists and botanical gardens. The work goes on with virtually no money to support it. In Madrid, Marta Chirino Argenta works tirelessly to produce line drawings for books such as *Flores Ibericos* published by the Real Jardín Botánico. Another of their artists, Juan Luis Castillo, has embraced modern technology by creating wonderful computer generated images of plants such as the orchid shown here. His view is that all art images originate in the artist's mind and travel from there to his hand regardless of whether it is pixels or paint which create the finished product. I think Bauer might have been tempted. He also creates exquisite ink drawings for scientific botanical illustration. Juan exhibits in America and elsewhere. He has received the Jill Smythies Award from the Linnean Society of London. Both Juan and Marta have received Gold Medals from the Royal Horticultural Society and Marta is a Member of the Society of Botanical Artists in the United Kingdom.

Another SBA Member, Reinhild Raistrick has recently written, illustrated and published a book on her trips to the Usambara Mountains of Tanzania in search of African Violet species. Millions of these plants in hybrid form are sold in Europe yet many of the species from which they are bred are fast disappearing in the wild. She

An original botanical illustration depicting *Myrmecophilla tibicinis* made by Juan Luis Castillo, who employs digital methods to achieve scientific accuracy in his botanic art.

Saintpaulia orbicularis by Reinhild Raistrick SBA from *African Violets*. This delicate species was painted in situ in 1991 in what was then a very vulnerable area of the West Usambara.

has received great support from the Royal Botanic Gardens, Kew, in her endeavours as this is not the sort of thing they can fund these days. Botanists have to take a camera but in this book you can enjoy paintings, not just of African violets but other beautiful flowers of the region. A number of Reinhild's paintings are already in the collection at Kew and one hopes the African Violets will join them.

At Chelsea Physic Garden work has gone on over a number of years in the creation of a florilegium using the talents of many of today's botanical artists. It is fitting that this should be done in a place which has an association with another great artist of the past, Georg Dionysius Ehret. One thing is certain—the link between artists and gardens will continue for as long as the pursuit of beauty and knowledge remain, for some, of greater importance than the acquisition of money.

1   Lack, Walter H, "A Garden for Eternity", *The Codex Liechtenstein,* Munich: Benteli Verlag, 2001.
2   North, Marianne, *A Vision of Eden: The Life and Work of Marianne North,* London: Holt, Rinehart and Winston, 1980.
3   North, Marianne, *A Vision of Eden: The Life and Work of Marianne North.*
4   North, Marianne, *A Vision of Eden: The Life and Work of Marianne North.*
5   Vellacott, H ed., *Some Recollections of a Happy Life: Marianne North in Australia and New Zealand,* Melbourne: Edward Arnold Australia, 1986.

DIAMOND BOTANICAL GARDENS AND WATERFALL

The Diamond Botanical Gardens are one of three botanic gardens on St Lucia. Although only established in the late 1980s, the Gardens' history reaches right back to the island's colonial past. The Gardens are sited on the Diamond and Soufriere estates, which are a portion of a 2,000 acre site that was bequeathed to three brothers with the surname Devaux by King Louis XIV in 1713. In 1784, the governor of St Lucia, Baron de Laborie, discovered a number of natural volcanic sulphur springs on one of the estates, and sent samples from them back to France for analysis. The minerals in the springs were found to be exceedingly rich, with great potential health benefits, and King Louis the XVI financed the construction of 12 baths for his troops to bathe in, fed by the spring. These baths were destroyed during the Brigands War, but were restored to their former glory in 1923.

20 years ago, Joan Devaux, a descendent of the original Devaux brothers, developed the site around the baths into botanical gardens. Although only six acres across, the Gardens are nonetheless very impressive. Nestling at the foot of a volcano, the Diamond River flows straight through the grounds, black with mineral sediment, and spills over a vast rockface in a magnificent waterfall to a lake below. A nature trail leads through the Gardens to a Red Cedar and Mahogany arboretum ending up at an original sugar mill that originally served the Soufriere estate.

The Gardens specialise in tropical plants, with a specific emphasis on banana palms (the banana industry in St Lucia is second only to the tourism industry in importance). With a broad selection of tropical trees and plants from around the world, the Gardens are unmistakably Carribean, with local shrubs, fruit trees and flowers confirming this. The historical baths are open to the public and there is a selection of mud and mineral springs that attract a healthy number of visitors every year.

tag: One of a variety of vibrantly-coloured plants hosted by the Gardens. above: The Diamond Waterfall, created by hot sulphur springs that emerge from underground. opposite page top left: A gazebo, which offers the Gardens' visitors a shady respite from which to enjoy the surrounding trees and flowers. opposite top right: View of one of the many meandering paths through the Gardens. opposite bottom: The fruit of a Nutmeg Tree (*Myristica fragrans*).

The Botanic Garden of Uppsala University is Sweden's foremost horticultural site and, like Kew or Padua, has always been committed to its role as an educational centre. It was founded in 1655 by the University's medicine professor, Olof Rudbeck, but the Garden was largely destroyed in a fire in the early eighteenth century. It was not until 1741, when eminent botanist Carolus Linnaeus took charge, that Uppsala properly found its feet. Linnaeus was responsible for collating an anthology of international plants, and methodically ordering and cultivating them. Using all his connections around the world, he gathered thousands of exotic plants, and turned the Garden into one of the foremost in the world. This period of prosperity was short-lived however—due to the swampy location of the Garden, next to the River Fyrisån, the grounds were hard to maintain. When King Gustaf III donated the Uppsala Castle Garden to the University in 1787, the Botanic Garden was moved there from its previous site, and Linnaeus' garden fell into disrepair.

Today the Botanic Garden occupies both sites, and extends over 34 acres. Linnaeus' landscape model has been replicated on its original site, and includes all the species that he is known to have cultivated based on his texts, diaries and visitors' accounts of his botanical research. His former home has been converted into the Linnaeus Museum, paying tribute to his contributions to botanical study. In the Aquarium palustre (the marsh pond), a species of twinflower, Linnea borealis, was named after him. The Baroque Garden has been restored according to the plan from the 1750s. Linneanum still houses an orangery, the oldest plants grown there are Linnaeus' laurels, four trees that are over 250 years old.

The main tasks for the Botanic Garden is to provide plant material and horticultural support for research and education within Uppsala University and to promote public awareness on all issues concerning biological diversity. Each year, more than 1,000 students are tutored in botany, pharmacology, horticulture or

ecology. The Garden regularly arranges guided tours, exhibitions and events open to the public.

Toward fulfilling its aims of plant scholarship, the Garden is arranged into categories, including economic plants, rock and arid gardens, stone troughs, peat beds, and annual beds. There is also a Tropical Greenhouse that contains over 4,000 species from the warmer climatic zones.

tag: A Hay-scented Orchid (*Dendrochilum glumaceum*), which is located in the Tropical Orchid House. previous pages left: The original plan of the Garden, 1745. previous pages middle: Carolus Linneaus. previous pages right: The Baroque Garden in winter. above: A small museum in the Garden, originally used to accommodate Carolus Linnaeus' collections. left: A *Pulmonaria maculata* which, like many other *Boraginaceae* species, changes colour when the it matures. opposite: The Baroque Garden, which dates back to 1660 and was remodelled in 1750. all photographs: Magnus Lidén/The Botanic Garden of Uppsala University.

SWITZERLAND
GENEVA CONSERVATORY AND BOTANIC GARDEN

The Geneva Conservatory and Botanic Garden was conceived in the early nineteenth century in the spirit of naturalism prevalent at the time. Established in the centre of the city, the Garden today occupies 46 acres of land along the lake. It holds an important collection of orchids, an arboretum, a historical rose garden, a small animal enclosure and various pedagogic and ornamental plantings. Several greenhouses supplement the collection with tropical and subtropical species.

A splendid and exhaustive alpine garden, called the "Alpinum" displays shrubs and plants from the alpine world. The plants are arranged according to geographical origin and are interspersed with rockeries and integrated water features that are directed into cascades and pools, evoking the magnificent mountains they originate from. They are planted in terraces of 'tuff' walls—walls built from naturally deposited limestone, indigenous to the alpine environment, which encourage plant growth due to their porous qualities.

At the heart of the Garden lies the Conservatory, which contains a state-of-the-art research laboratory, a botanical library possessing over 220,000 volumes, and one of the world's largest herbariums—containing no less than 5.5 million species.

The Garden is one of the most prominent in terms of its scientific contributions, and it is leading the field in terms of research and analysis of plant-life, specifically in the specialist areas of floristics, biosystematics and morphology. In addition to its scientific excellence, the Conservatory operates several interesting and engaging public programmes and features; the Garden of Scent and Touch is one such unique endeavour. Built in 1990 in collaboration with several Swiss malvoyants or visual disability organisations,

the adaptation and arrangement of this garden is concerned with tactile qualities, perfume and the immediacy of plant shape. Exhibits are annotated both at floor level and in Braille and some display units carry perfume vials to assist their enjoyment.

Another unusual public programme is the Welcome Class Initiative, an outreach programme for immigrant children. This

makes use of the Garden's facilities to encourage the children to feel integrated in the local culture and environment, and a safe, enjoyable environment in which to learn French. Participants are encouraged to share their own cultural heritage, and the backdrop of botany and flora is utilised as a bridge for social integration. These initiatives demonstrate the Conservatory's aspirations and capacity for social as well as scientific impact.

tag: One of the Garden's many bedded plants. above: The Rhododendron Alley. opposite top: The Old Botanical Conservatory. opposite bottom: View of the Temperate Greenhouse. all photographs: CJB D Roguet, B Renaud and H Gander.

TURKEY
NEZAHAT GÖKYIGIT BOTANICAL GARDEN, ISTANBUL

This uniquely situated Garden sits on a busy motorway intersection in a residential area of Istanbul. It was established in 1995 by Mr A Nihat Gökyigit as a memorial park for his late wife Nezahat. It is the first and only botanic garden in the world to be situated on a motorway junction, and it has increased the green space in the city by over 17 per cent, providing a much needed 'lung' for this bustling metropolis.

The Garden takes its re-greening priority very seriously, a fact which is reflected in the sheer speed of its establishment and growth. Over the past decade over 50,000 trees and shrubs from 150 different native or alien species have been planted, although not all have survived the harsh conditions of the Garden. A planting database has been set up in order to record and categorise the Garden's extensive collection of more than 250 geophyte species from 36 genera. In an extension of this eco-friendly policy, recycling is seen as an important aspect of the Garden and a composting system has been established to obtain

natural organic fertilizer to improve soil fertility. Disused railway sleepers have been given a second life as sleepers for a collection of over 250 geophyte species.

The Garden contains within it an adminsitration building, a library and a herbarium, as well as an education centre. In a city with very few horticutural resources, this latter facility plays an essential role. It provides school courses for children and a venue for lectures, workshop meetings and training courses for school teachers, adults, staff and volunteers. During the summer months the Garden invites a number of students to come and work at the park as part of their practical experience and also hosts participants from various international congresses. In addition to these activities, the Children's Garden provides an opportunity for younger visitors to learn how to grow and care for flowers and vegetables.

One of the key features of the Garden is a viewing gazebo, from which one can survey not only the landscape itself but their situation within their urban envrions. A waterfall at the foot of the gazebo runs into two pools, enhancing the vista. A more recent addition is Erturul Island, which is an extension to this decidedly urban botanic garden, which can be accessed through a tunnel under the motorway. It is currently being landscaped, with planting schemes along the trails devised for displaying native species. Many cherry trees, gifts to the Garden from Japan, have been planted, and a medicinal and aromatic plants area is under development. In this area an ancient olive tree, which traditionally symbolises peace has been transplanted near to an amphitheatre. In a continuation of the environmental concerns and to raise awareness of the increasing problems of water scarcity, an area is devoted to plants which will tolerate arid conditions and which may prove useful in the global struggle against soil erosion and desertification.

tag: An *Iris paradoxa*. photograph: Margaret Johnson. left: Aerial view of the Administration Building, Library, Herbarium and Greenhouse. photograph: Kemal Öner. opposite top: The Rock Garden. photograph: Margaret Johnson. opposite bottom: A Rose (*Rosa*) and Vine (*Trachelospermum*) pergola.

BIRMINGHAM BOTANICAL GARDENS

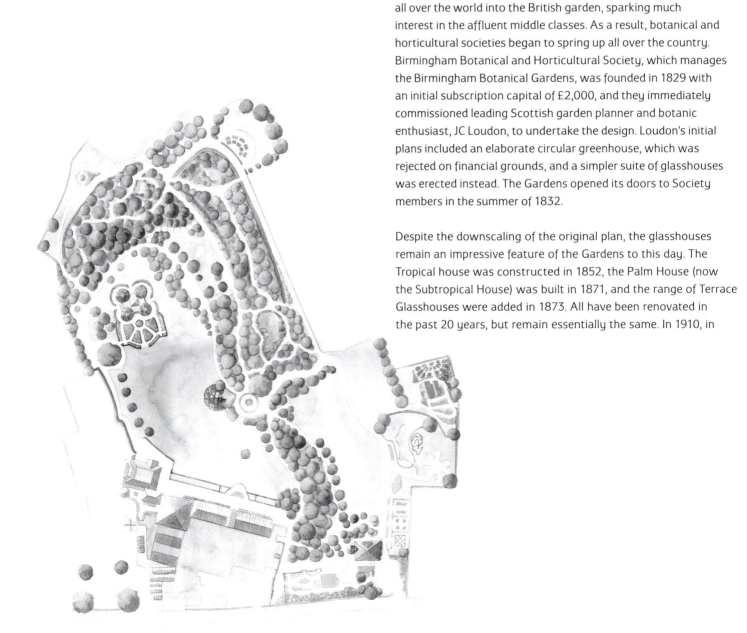

During the colonial era of the eighteenth and early nineteenth centuries, global exploration brought many new plants from all over the world into the British garden, sparking much interest in the affluent middle classes. As a result, botanical and horticultural societies began to spring up all over the country. Birmingham Botanical and Horticultural Society, which manages the Birmingham Botanical Gardens, was founded in 1829 with an initial subscription capital of £2,000, and they immediately commissioned leading Scottish garden planner and botanic enthusiast, JC Loudon, to undertake the design. Loudon's initial plans included an elaborate circular greenhouse, which was rejected on financial grounds, and a simpler suite of glasshouses was erected instead. The Gardens opened its doors to Society members in the summer of 1832.

Despite the downscaling of the original plan, the glasshouses remain an impressive feature of the Gardens to this day. The Tropical house was constructed in 1852, the Palm House (now the Subtropical House) was built in 1871, and the range of Terrace Glasshouses were added in 1873. All have been renovated in the past 20 years, but remain essentially the same. In 1910, in

an effort to increase the Society's membership, a zoological collection was also established—well before any other such enterprise in the region. Over the subsequent years, seals, alligators, pythons, wallabies and bears were all housed on the grounds. In recent years an aviary has been constructed, with four sections housing a variety of birds.

The rest of the Gardens remain much according to Loudon's original plans, offering a mosaic of grounds in contrasting styles to appeal to a variety of interests. A small plot devoted to British native plants recreates a flower-rich hay meadow, of the kind that used to be so common in England. Other units have been modelled as illustrations of horticulture throughout the ages. The glasshouses and large parts of the Gardens have been devoted to collections representing the plants of various geographical regions around the world. Paths bisect these sections, named after explorers, planners and authors who have made outstanding contributions to botany. Conservation is highly prioritised at Birmingham Botanical Gardens, and to this end, an education programme has been developed over the years, which has proved increasingly popular with schools and which, it is hoped, will help

to disseminate information about plants and promote an attitude of respect for the world's vegetable resources.

tag: Fowers in the Rhododendron and Azalea Walk. opposite: A contemporary plan of the Gardens. above: The Gardens' Bandstand in summer. below: The Lawn Aviary in summer.

Most botanic gardens foster some kind of relationship with academia, encouraging scientific research into various aspects of horticulture. The Italian gardens in Padua, Pisa and Florence initiated this trend, and it is often fundamental to the function (and funding) of a botanic garden. Very few gardens, however, can trace the history of how this academic study was conducted and how botany itself has developed as a scientific practise. In this sense, the Cambridge University Botanic Garden is unique.

The Cambridge University Botanic Garden is divided in to roughly two areas: the older section to the west, established by John Stevens Henslow in 1846, and the newer section, which was added in the 1950s. The older, Victorian section represents the era's fascination with the sourcing of plants from all corners of the British Empire, and their subsequent systemisation and classification. During this period, botanic gardens represented science's new ability to temper and master Mother Nature; also to create emblematic and awe inspiring landscapes celebrating the curators' horticultural skill. This is probably most apparent in the Systematic Beds, in which a wide range of species are cultivated according to their taxonomic group. This Victorian section at Cambridge is not so much an ode to nature, as an ode to botany itself.

Moving on to the newer, mid-twentieth century gardens, one is immediately aware of a shift in botanical practise. The plants here are arranged thematically, and hint at the new scientific developments in botany regarding "our understanding of interactions between different species", as the curators themselves state. The Autumn Garden, Scented Garden, and Dry Garden are the highlights of this section. The latter features

predominantly Mediterranean plants (though there are some British natives too) that require absolutely no watering. There is also the fairly small Chronological Bed, which contains plants from around the globe arranged into a timeline based on their introduction into Britain. Starting in Roman times, and ending in the eighteenth century, this section also covertly tracks the trajectory of British colonial relationships.

Cambridge University Botanic Garden is not merely historical evidence of botanical study, it is also a very important academic centre for the geneticists, plant scientists, zoologists, and geographers at the University. The Garden contains a whole series of structures built for the express purpose of scientific research, including laboratories, glasshouses and growth rooms. The design of the landscape has also come under scrutiny in the past by architecture students. With one foot in the past and one in the future, this garden is one of the world leaders in the field of scientific research and development.

tag: King Protea (*Protea cynaroides*), situated in the Temperate House. previous pages left: Historical depiction of the Garden's Succulent House. previous pages middle: The Rock Garden, which overlooks the Garden's lake. previous pages right: The Garden in summer. above: The Garden in summer. below: The Greenhouse with the Garden's landscaped summer beds in the foreground. opposite: A pathway meanders through the Garden in spring.

UNITED KINGDOM
CHELSEA PHYSIC GARDEN

The exclusive neighbourhood of Chelsea in London is often associated with its famous annual flower show, however, the horticultural history of the area goes back far beyond this event. The Chelsea Physic Garden, one of the hidden gems of London, was established in 1673 by the Society of Apothecaries, as a resource for physicians to train their apprentices in the identification of medicinal plants. The location was chosen for its proximity to the Thames, which provided a slightly warmer climate, allowing a broader range of non-native plants to be cultivated, and also enabled plants from overseas to be easily offloaded.

In the mid-eighteenth century, the Garden was purchased by Dr Hans Sloane, although the management remained in the hands of the Society of Apothecaries. Under Sloanes ownership, environments for supporting different types of plants were built, including the Pond Rock Garden, which built in 1773 is the oldest rock garden in England.

Britain's colonial ventures contributed greatly to the collections at Chelsea Physic Garden, and many exotic species continue to thrive in its grounds. Its collection of native Chinese specimens is especially extensive, with two Madenhair trees standing near the entrance, and Ehretia Dicksonii whose yellow berries border the main lawn. The tea plants (*Camellia sinensis*) were brought over in 1848 by Robert Fortune, who used Wardian cases—a kind of miniature greenhouse—to safely transport the seedlings from China, and it was these seedlings that became the basis for the tea industry in India. The mild microclimate in the Garden allows many tender plants from the Mediterranean region and the Canary Islands to be grown. The glasshouses further contain a collection of tropical, sub-tropical and Mediterranean species.

Although older than nearby Kew Gardens, Chelsea only opened to the public (and still only on a limited basis) in 1983. It is possible, however, for the public to catch up on the 400 year history of the Garden in displays of plants that were introduced by people associated with the Garden's history. Chelsea Physic Garden today holds approximately 5,000 taxa, with an emphasis on medicinal

plants and those of ethnobotanical interest, as well as rare and endangered species. In recent years The Chelsea Physic Garden has developed a major role in public education focusing on the renewed interest in natural medicine.

tag: Flowerheads of the *Onopordum cyprium* plant. top: The statue of Sir Hans Sloane, the Garden's benefactor. The carts on either side represent his connection with Carolus Linnaeus. bottom: A Judas Tree (*Cercis siliquastrum*), situated next to the Garden of World Medicine. opposite: View of the Culinary Beds, with the Pharmaceutical Beds in the background.

Architecturally, horticulturally, socially: The Eden Project is spectacular on all counts. Perhaps the only major botanic institution to be established in the twenty-first century, it has already become one of the most significant and most flamboyant botanic gardens in the world.

The Eden Project was conceived first and foremost as a site where the public could reconnect with its local and global environment. It highly values the bonds between plants, people and places, and this municipal concern is played out under the curvaceous, amoebic glass domes of the site in Cornwall, one of Britain's most beautiful and most historical territories.

The Eden Project was the brainchild of Tim Smit, a Netherlands-born British businessman. West Country horticulture owes a lot to this man: Smit was also responsible for the extensive redevelopment and regeneration of the Lost Gardens of Heligan, near Mevagissey in Cornwall. In some ways, Eden is just as much an architectural venture as a horticultural one, and Smit worked closely with architect Nicholas Grimshaw (now Sir Nicholas) to bring the Eden idea to physical reality. Grimshaw's primary vision for the site was the inclusion of the two transparent domes,

and he based their iconic design on the undulating shape of the biome houseplant. He was also responsible for designing what is known as The Core, a late addition to the Eden site and home to the educational resource centre. Grimshaw developed the geometrical design of The Core's roof in collaboration with sculptor Peter Randall-Page and Mike Purvis of structural engineers SKM Anthony Hunts. The design was based on the Phyllotaxis, the mathematical basis for nearly all plant growth; the 'opposing spirals' found in many plants such as the seeds in a sunflower's head, pinecones and pineapples. Throughout Eden's formation, a cross-fertilisation of architecture, ecology and botany is apparent.

The two 'biomes' house the two halves of Eden's botanic collection. The larger, the Humid Tropics Biome, is for tropical plants such as fruiting banana trees, coffee, rubber and giant bamboo, and is kept at a tropical temperature. The smaller of the two, the Warm Temperate Biome, houses familiar warm temperate and arid plants such as olives and grape vines. In both cases, the architecture is just as intriguing as the plant life. The biomes are constructed from a tubular steel frame with mostly hexagonal transparent panels made from a complex plastic knows as ETFE. The structure is completely self-supporting, with no internal supports, and takes the form of a geodesic construction: The influence of the glass Climatron building at the Missouri Botanical Gardens, built in 1960, should not be underestimated here. The plastic panes vary in size up to nine metres wide, with the largest at the top of the structure.

The use of plastic in the architectural design signals the Eden Project's central environmental concerns. Throughout the Gardens, there are attractions and information signs on Global Warming, and why plants are so important to our way of life, with one display showing what our world would be like without plant life. The massive amounts of water required to create the humid conditions of the Tropical Biome come from sanitised rainwater, and Eden uses Green Tariff Electricity, generated by

one of the many wind turbines in Cornwall. It is also home to the Eden Trust, a registered charity devoted to the promotion of the vital relationship between plants and people, leading towards a sustainable future for all. The Garden is passionately concerned with the advancement of renewable energy, biodegradable waste and effluent control systems.

The Eden Project is unique in that it raises the question of what concerns a contemporary botanic garden should have, be it recreation, education or conservation. Regardless of the modern form it takes, the Eden Project has a desire to return to nature at the heart of its operations. Built with the luxuries of modernity, the Eden Project has forged a new ecological role for botanic gardens and, since its opening in 2001, has seen many international gardens follow suit.

tag: The Mosaic path designed by Elaina Goodwin, one of the many artworks exhibited at the Eden project. previous pages left: The Project's Warm and Humid Biome. previous pages right: Overview of the Project's tesselated biomes, and the dual aspect skylights and solar panels on the roof in the foreground. below: Panoramic view of the Project. Opposite: A nighttime concert illuminates one of the Project's many biomes.

UNITED KINGDOM
ROYAL BOTANIC GARDENS, KEW

The Royal Botanical Gardens at Kew is one of the most famous and respected botanic gardens in the world. Recently awarded UNESCO World Heritage Site status, it covers 300 acres of land and 250 years of botanic heritage. Its collections include no less than 29,000 taxa and 110,000 living plant specimens. At this size it is one of the largest collections in the world.

The history of the Gardens dates back to the mid-seventeenth century. It was first recorded as an estate owned by the Capel family, who had a passion for horticulture. The Dutch House that still stands on the grounds today was built in 1631 by a Flemish merchant—its distinctive brickwork and rounded gables clearly reflecting Dutch architecture of the time. Over the next two

centuries, the site of Kew Park, or Kew Field, as it was variously known, passed through phases of political and monarchic tumult, gaining additions and structural embellishments along the way. Its picturesque surrounds on the bank of the Thames allowed it to evolve into a pleasure garden for eighteenth century aristocracy, who indulged their colonial interests by investing in architectural follies in various global styles. An example of this is the Grand Pagoda, erected in 1762. The structure rises to fifty feet at its peak and is tiered into ten octagonal levels. It has survived the centuries well, and opened to the public in 2007, providing visitors with a 360 degree panorama not only of the beautifully tended gardens but also the city of London stretching beyond.

In the early nineteenth century, Kew was purchased by King George III as a nursery for the younger royals in 1840 it was finally officially appropriated as a national botanic garden, and enlarged substantially. Several of the structures built since that time are noteworthy, both for their aesthetic properties and for the cutting edge technologies that were employed in their construction. This is particularly true for the glasshouses. The Palm House, for example, was erected in 1848 by architect Decimus Burton, and features the first large-scale structural use of wrought iron. The Temperate House, which is twice as large, was built in the latter half of the nineteenth century, and is the largest Victorian glasshouse in existence. And the recently added Princess of Wales Glasshouse is large enough to accommodate no fewer than ten microclimates.

The old Orangery, which now houses a cafe and restaurant, is another of the Gardens' elegant historic buildings. It originally boasted what must have been one of the earliest underfloor heating systems in horticultural use. In 1769 Sir John Parnell proclaimed it "filled completely, chiefly with oranges which bear exceptionally well and large". In stark contrast to its classical

proportioned facade, the sculptural White Peaks is a strikingly modern building, which features an exhibition space housing ever-changing displays of contemporary art.

The botanical displays are too numerous to summarise individually; amongst the outdoor ornamental exhibits, notable are the divinely fragrant Lilac Garden, which features in the old music hall number "Come down to Kew in Lilac Time", and the stunning panoramic Colour Spectrum Garden. The Secluded Garden is designed for sensory enjoyment, featuring a rippling water feature, and an array of rustling, fragrant planting. The Climbers and Creepers Section is one of the most recent outdoor exhibits and is designed for interactivity, encouraging young children to physically engage and play with the plants displayed.

Aside from its impressive architecture, landscaping, and collections, which attract over one million visitors a year, the Royal Botanical Gardens at Kew is one of the foremost scientific institutions of its kind in the world. Once referred to as the 'international metropolis of plants', it is respected worldwide for its outstanding living collection; it is also renowned for its world-class Herbarium as well as its scientific expertise and authority in plant diversity, conservation and sustainable development in the United Kingdom and around the world.

tag: The ten tiered Grand Pagoda, erected in 1762. previous pages left: Historical depiction of the Gardens' lake and the Grand Pagoda. previous pages middle and above: Kew Palace, once a nursey to young royals during the reign of King George III, is the Gardens' oldest building. previous pages right: The Palm House, which dates from the mid-nineteenth century. above: The Temperate House, the largest Victorian Greenhouse in existence. above: The Princess of Wales Glasshouse, which maintains ten microclimates in order to sustain a huge variety of plant life. all photographs: Kew Press.

UNITED KINGDOM
ROYAL BOTANIC GARDEN, EDINBURGH

The Royal Botanic Garden Edinburgh was founded in the seventeenth century as a physic garden, specifically growing medicinal plants for the benefit of the local resident and medical population. This first garden was in St Anne's Yard, part of the Holyrood Palace grounds, and occupied an area the size of a tennis court. It now extends to four sites—Edinburgh, Benmore (near Dunoon in Argyll), Logan (near Stranraer in Galloway), and Dawyck (near Peebles in the Borders), and has the one of the richest collections of plant species in the world.

The Garden is first and foremost a scientific institution, dedicated to discovering and describing plants and their relationships, evolution, conservation and biology. This research is underpinned by the Garden's internationally important collections of living and preserved plants, a large specialist library, and by modern well-equipped laboratories. It has been an important resource for the Scottish scientific community, as well as providing a beautiful landscape for residents and visitors to enjoy remarkable species of flora from around the world.

The Palm House—the tallest of its kind in the United Kingdom—houses a remarkable selection of tall plants. The Glasshouses provide an opportunity to see plants from all ten major climate zones, with imposing species being accommodated within the structure's radical design. The oldest plants are carefully tended in the Orchid and Cycad House, while the Tropical Aquatic House holds all the rainforest based plants. The Peat and Rock Houses contain the largest collection of Vieya Rhododendrons cultivated directly from New Guinea and Borneo—a truly spectacular sight.

In 1997, the Chinese Hillside opened to display Chinese plants in the largest collection of native Far Eastern plants in cultivation. Collected in the early years of the twentieth century, they are set within a beautiful and calming setting complete with a pool of water to idle beside while taking in the sights. The Rock Garden displays rare stones acquired from Arctic climes, with a collection of low-lying arctic plants that compliment the setting perfectly. Additional to these attractions, the Garden also boast a Scottish Health Garden and a Woodland Garden.

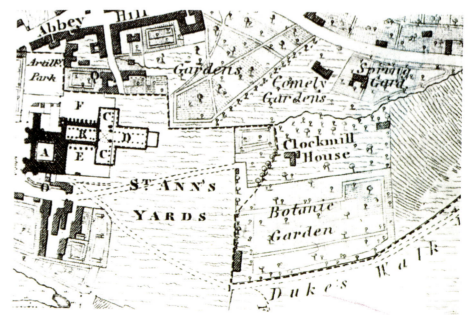

The Royal Botanic Garden of Edinburgh's Herbarium was set up in 1836. Herbariums provide a valuable record of plant life, and are an indispensable tool for botanists and hobbyists alike. A good herbarium has many specimens of each kind of plant marked up according to the location in which the specimen was found. This provides an indication of the botanical history and development of various regions. If the specimens are protected from pests and moisture, specimens will last for hundreds of years, indeed the oldest housed by Edinburgh dates back to 1697.

tag: Waterlilies in one of the Garden's ponds. opposite left: An eighteenth century engraving depicting the four gardens of The Royal Botanical Gardens. opposite right: An early plan maps out the site of the first incarnation of the Gardens. top right: detail of the wrought iron work, which can be found in the Palm House. top left: One of the Gardens' Greenhouses. below right: View overlooking the lake ands its plentiful plants and wildlife.

UNITED KINGDOM
UNIVERSITY OF OXFORD BOTANIC GARDEN

The University of Oxford Botanic Garden is the oldest botanic garden in the United Kingdom. Established in 1621, the Garden is at the heart of the University and the city of Oxford. It was established, in part, thanks to a handsome donation of £5,000 (equivalent to £3.5 million today), by the Earl of Danby, Sir Henry Danvers. He made clear the Garden should be intended for "the glorification of God and for the furtherance of learning", and the institution is still committed to these ideals, though its ecclesiastical aims have slowly morphed into those glorifying nature.

Order and geometry is common to any British botanical garden, and this is nowhere more evident than in Oxford's largest section, the Walled Garden. Planned and designed by Sir Isaac Bayley Balfour in 1884, this area is made up of numerous sets of rectangular borders. The plants within each sub-section are arranged according to their familial classification, including all the different genera that still share enough basic similarities to suggest that they have all evolved from a common ancestor. The Glasshouses at Oxford are less regimented. Alongside 100-year-old cacti and extravagant lilies, cocoa plants and sugarcane grow, as well as orange and banana trees.

The first catalogue of the collection was published in 1648. In 1992 Henry Scholick bequeathed his copy of the catalogue to the Garden. To mark this generous gift the curators of the Garden meticulously researched the contents of the catalogue and recreated the small garden Scholick described. Interestingly, many of the species that thrive in this recreated garden are still common plants in domestic English gardens.

Today, the Garden encompasses various specialised sections including the big Herbaceous Border, the Rock Garden, the Spring Walk, the Water Garden and the Bog Garden. Many of these features are recent developments, and add a contemporary horticultural element to the historic features of the Walled Garden they surround.

The oldest specimen in the collection is that staple of any English garden, the yew. In 1642, the Garden's first curator, Jacob Bobart, planted a whole series of Paired Yews (*Taxus Baccata*) and would clip them into various exotic shapes, rarely seen in gardens of the time. Now only one lonely yew remains, a reminder of the University of Oxford Botanic Garden's grand beginnings over 400 years ago.

tag and previous pages bottom right: A gravel pathway leading through one of the Gardens many herbaceous beds. previous page left: Illustration from *Alice in Wonderland* depicting Alice and the Queen of Hearts in the Garden. previous pages top left: The Bog Garden with Magdalen College in the background. previous pages top right: The Garden in winter. previous pages bottom left: A stone urn covered in frost during winter in the Garden. opposite: Herbaceous border planted with *Gladiolus communis byzantinus* and *Polygonum bistorta superbum*. right: The Garden in autumn. below: Magdalen College with summer poppies in the foreground. All photographs: Clive Nichols/ Oxford Botanic Garden.

ALASKA BOTANICAL GARDEN, ANCHORAGE

The Alaska Botanical Garden has a large range of annuals and perennials specific to south-central Alaska. Famous for the blue poppy, the gardens also include 150 varieties of plants native to the region and 1,100 other varieties of hardy perennials, culminating in an impressive showcase of the summer arctic growing season. Paths have been steadily integrated within the landscape throughout its history to provide easy walking among the ten acres of native spruce and birch forest. There are fertile flowerbeds that interweave and disperse according to the route taken along the trails, where interpretive signs guide visitors and identify plants in the Rock Garden, Herb Garden, perennial gardens, and along a wildflower walk. The winding nature trail offers a view of the Chugach Mountains, and leads the visitor through a natural wetland.

Planning for the Alaska Botanical Garden began in 1983 when members of the national Horticultural Association considered the creation of an arboretum and its benefits for the land, people and economy. In 1986 a planning committee was formed and the Garden was officially constituted as a non-profit organisation with the aim of creating a public garden for the purposes of education and pleasure. In 1990 the municipality of Anchorage approved the proposed Masterplan, and the initial land use agreement for 110 acres adjacent to Campbell Airstrip Road was completed. Ten years after the initial planning, the first gardens were planted and the Garden's grand opening took place on 25 July 1993.

The Athabascan Indians once used the site where the garden now stands. Via negotiation and through the efforts of the Horticultural Association, claims of title to the land passed from federal to municipal administration, allowing for a flexible working programme for the area in collaboration with the local people. In 2003 an agreement was reached between the garden and the municipality of Anchorage allowing for further development of the land. This resulted in a lease being signed for 80 acres of land on the east side of Anchorage. Much of this proposed expansion is still undergoing development, but the plans promise an extension of the current natural preservation area.

The growing season is about 120 days long in the Anchorage area with the first frost-free day occurring around the end of April, with the first frost returning in the middle of September. Changes in elevation of 500 feet in the Anchorage 'bowl' significantly effect the growing season, meaning that depending on weather conditions, the speed of growth for most plants is reduced or accelerated significantly. The Alaska Botanical Garden does not shy away from the challenges of a reduced growing season or changeable elevations, and succeeds in cultivating a range of native and non-native plants in a rare collection of arctic flora. Visitors also benefit from the Garden's healthy relationship with the local wildlife, indeed it is not uncommon to see a moose or bear strolling through the lanes that carve out the Garden's unique grounds.

tag: This Columbine (*Aquilegia*), along with many other varieties, grace the beds at Alaska Botanic Garden. opposite top left: Flowers in the Demo Garden. opposite top right: the Himalayan Blue Poppy (*Meconopsis lingholm*) which only grows in a cool summer climate. opposite bottom: View of the Herb Garden.

UNITED STATES
TUCSON BOTANICAL GARDENS, ARIZONA

Tucked away in the heart of the city, Tucson Botanical Gardens is a small garden that focuses on its local community. It was initially founded in 1964 by horticulture enthusiast, Harrison G Yocum, who opened his own garden to the public, exhibiting his extensive collection of cacti, palms and other native plants. The grounds were fairly small, and the organisation that managed the property, the Tusson Botanical Garden Organisation, was on the lookout for a more substantial home for the Garden. In 1975, the answer to their prayers came in the form of a substantial land donation from one Mrs Bernice Porter. Her house still stands today, dating back to the 1920s and constructed of adobe bricks made on site. The Gardens have since expanded to cover six acres of land, divided into 16 landscaped gardens.

Rather than focusing on the pursuit of scientific research, these Gardens seek to hold together the strands of local community, cultural heritage and environmental stewardship. The history of the area is commemorated in the grounds and the collections, with a specific emphasis on the background and customs of the Tohono O'dharn Indians, that initially populated the area. Seed varieties have been preserved from this pre-colonial time, teaching visitors about the native plants used by the Native Americans for food, construction and medicine, as well as providing scientists and historians with crucial information about the farming methods and daily life of the Tohono O'dharns. The Gardens further chart the evolution of gardening in Tucson and the south western states, documenting, for example, the period in the 1930s, when landowners would build fanciful gardens inspired by tales of the Orient.

The original garden of the Porter family is also preserved, designed in the barrio-garden style of Tucson's Mexican-American community. Further gardens include the Backyard Bird Garden, which shows visitors how to attract local bird species into their gardens, the Butterfly Garden, featuring plants that attract butterflies, and a Cactus and Succulent Garden.

Environmental concern is a big theme of the Tucson Botanical Gardens, and many of the exhibits, such as the Dr Raymond C Allen Memorial Iris Garden, feature low water-use plants. The Xeriscape Demonstration Garden shows visitors how it is possible to maintain style and responsibility in their own home garden. In this way the Gardens' focus is on the community of Tucson as it was in the past, and ways in which it can prosper in the future.

tag: A Saguaro Cactus (*Carnegiea gigantean*). top: The Children's Garden, which aims to encourage young people's interest in botany. bottom: The Sensory Patios, which are organised to stimulate smell, sight, sound, touch and taste. opposite: The Desert Section, which includes several spiky agaves and arid plants.

UNITED STATES
MAGNOLIA PLANTATION AND GARDENS, CHARLESTON

The Charleston Magnolia Plantation in South Carolina, has a long history that runs parallel to that of the American South. It began as a plot of land, bequeathed to Thomas Drayton and his wife, Ann Fox, on the occasion of their marriage, and has remained in the family ever since. South Carolina was one of the major junctions of the slave trade, and Drayton, amongst many others, was responsible for the ferrying of Africans from the West Indies to North America. The Plantation was cultivated and developed by slaves over the subsequent 200 years. Today the Plantation acknowledges this history, by investing in extensive research into the role of slaves on the estate and the surrounding local community, and commemorating this role in events, programmes and educational centres. The impressive Barbados Tropical Garden is also, to some degree, a tribute to the chequered history of the area.

Since the days of Thomas Drayton, the Charleston Magnolia Plantation has changed extensively—houses have come and gone and the collection has grown in scope and orientation. One thing that has remained the same, however, is a Live Oak (*Quercus virginiana*), planted by Drayton on the corner of Bridge Square, its arms spreading over the lawn. Local folklore claims that the ghost of the recently deceased J Drayton Hastie (died December 2002), whose ashes were placed in the hollow of the tree, is occasionally visible in the mighty trunk of Charleston's oldest inhabitant.

The collection at Magnolia is extensive, with a specific emphasis on Camellias. John Grimke Drayton planted the first *Camellia japonica* in the 1820s, and by the 1970s, nearly 900 varieties of the plant were recorded in the Gardens, of which approximately 150 were bred in the nurseries there. Charleston has even constructed a replica of the maze at Hampton Court, made with Camellia shrubs instead of boxwood.

This maze is one example of the Plantation's rather inventive take on the curation of the Gardens, though there are many more. Besides the Swamp and Snapdragon Gardens, there is also the Biblical Garden. Split into two sections—Old and New Testament—the Gardens features plants named throughout the Bible, including those that have symbolic ties with religious tales. The Gardens is said to be devoted to the numerous churchgoers involved in Magnolia's history, and again establishes the extent to which the estate is tied to its own history.

tag: Atamasco lilies. left: An Iris in bloom of the bank of the Ashley River. opposite top right: Ferns and cypresses, a variety of which grow on the Plantation. opposite top left: A carpet of Magnolia Soulangiana petals, which line the ground. opposite bottom: The Plantation's Bamboo Pond.

The Fairchild Tropical Botanic Garden is located in Coral Gables (a city within Miami-Dade County) and was opened to the public in 1938. The Garden possesses a variety of palms, fruit trees, cycads, vines and flowering trees, the majority of which were collected from the wild, making it the only tropical botanic garden in mainland America.

The Garden was founded by Robert H Montgomery, a successful businessman and dedicated collector of plants, whose dream was to create a location in the United States where tropical plants could grow outdoors all year round. This dream was realised through his collaboration with David Fairchild, one of the most prolific plant explorers of all time, who provided extensive knowledge and advice to the project and whom Montgomery subsequently named his garden after.

The Garden spans 83 acres with its primary features established over the first 15 years of its life. The Palmetum, Bailey Palm Glade, Vine Pergola, Amphitheatre, Library and Museum, 14 lakes, irrigation systems and the Garden House Auditorium ensure that the Garden has the variety necessary to appeal to almost any visitor. Later additions included the Hawkes Laboratory, in 1960, and the Corbin Education Building in 1972, providing the Garden with the research infrastructure necessary to continue its well-respected botanical programme. The Garden also contains a number of specialist attractions, such as the Simons Rainforest—an outdoor collection of vegetation traditionally found in rainforests across the globe—and the Whitman Tropical Fruit Pavilion, which aims to continue David Fairchild's legacy of the collection and conservation of tropical fruit.

The emphasis that the Fairchild Tropical Botanic Garden places upon research, conservation and education is reflected in both these projects and more specifically in its status as a world leader in palm research. The Garden reports more than 550 identified palm species, together with the largest number palm taxa of any botanic garden in the world. It also boasts a large number of endangered, threatened or rare species from South Florida and the Puerto Rican archipelago, assisting in their conservation.

tag: *Costus barbatus*. photograph: Lorena Alban/ FTBG. opposite left: *White Tower,* 1997. photograph: Lorena Alban/ FTBG. opposite right: *Roystonea regia*. photograph: Lorena Alban/FTBG. below: The Garden at sunrise. photograph: Gaby Orihuela/FTBG.

Joyce and Ed Doty retired to the Hawaiian island of Kauaì in 1982, bought a large property and began to landscape their front garden. That small undertaking has now developed into a massive 240 acre botanic garden, complete with one of the largest collections of bronze sculptures in America. In 1999 the Dotys created a not-for-profit foundation, to which they donated the entire Gardens and opened their doors to the public. Na Àina Kai's gardens showcase numerous natural and designed water features, unique gazebos, arbors, bridges and benches— all envisioned and built by the Dotys. In addition to the sculpture collection, the gardens include a hedge maze, a koi-filled lagoon, waterfalls, a forest of 60,000 hardwood trees, miles of trails, and a beautiful and secluded white sand beach.

The Primary Garden (which was a cattle pasture when the Dotys first purchased the property), is home to a collection of palms, a desert garden of cacti, succulents and native Hawaiian plants, and a herb and spice garden. It also accommodates Shower Tree Park, whose flowering trees are draped with staghorn ferns and orchids.

Winding through one of the least tamed areas of Na Àina Kai, the Wild Garden is a gently sloping path descending to the ocean alongside Kuliha'ili Stream. Here, beneath the shadowy rainforest canopy, ferns, bamboo and spongy mosses grow abundantly. heliconias, gingers, noni, arnotto, and cycads flourish alongside breadfruit, jackfruit and cashew trees, and spices such as cinnamon and cardamom. The wild forest abounds with indigenous and exotic song-birds such as the White-rumped Shama.

On Kilohana Plantation, more than 60,000 thriving tropical hardwood trees, all raised from seed on the property, promise to one day provide a replenishing source of timber for established international markets. This forest primarily comprises teak trees, but contains 20 other hardwood species including mahoganies, zebrawood and rosewood, and covers 110 acres of land.

Na Àina Kai's Under the Rainbow Children's Garden features a 16 foot tall bronze sculpture/fountain depicting Jack and the Beanstalk, a maze in the shape of a gecko, a rubbertree treehouse, a jungle area with bridges, tunnels and slides, and a train.

The most recently developed area of the Garden is a recreation of the pie-slice land division between the mountains and the ocean, known as Ahupua'a. A dozen bronze sculptures illustrate traditional crafts and activities, performed by the indigenous peoples of the island including kapa-making, mat-weaving, fishing, net-mending, canoe-making, planting and hunting. The mountain and perimeter plantings are all native Hawaiian or Polynesian plants.

tag: The Poinciana Maze, which is comprised by 2,400 Mock Orange plants (*Murraya paniculata*). photograph: Staff Siegel. opposite top left: The Ka'ula Lagoon in the Primary Garden. The sculpture/fountain is entitled Flight of the Tropic Birds by the artist Pancho Vinning. photograph: Na Àina Kai Botanical Garden/Joanne Smith. opposite top right: The Ka'ula Lagoon in the Primary Garden. In the background is the Japanese Teahouse. photograph: Na Àina Kai Botanical Garden/Joanne Smith. opposite bottom left: View of the Kaluakai Meadows, and the ocean, from the Wild Garden. photograph: Na Àina Kai Botanical Garden/Joanne Smith. opposite bottom right: Rows of Big Leaf Mahogany (*Swietenia macrophylla*). These hardwood trees are among the more than 60,000 hardwoods growing on the property. photograph: Na Àina Kai Botanical Garden/Joanne Smith.

Not so much a botanic garden as a sprawling cultural complex, The Huntington Botanical Gardens is a centre serving all manner of municipal and public needs. Incorporating the Library, numerous galleries and the botanical collection, Huntington was founded by railroad and real estate magnate Henry Edwards Huntington in 1919. Based in his Georgian mansion, the centre was established primarily as a social attraction, as a public home to a diverse and disparate collection of literal and artistic works. Despite the Gardens' current status and eminence, it seems that its inception was almost accidental. The development of the site has been protracted, yet thorough, with the landscape taking its time to mature into the charming grounds they are today. Originally the grounds were an addendum to the prominent Library and galleries, with the Gardens created without the determinedly didactic goals that most are born from. When Henry Huntington purchased the site in 1903, he commissioned his superintendent, William Hertrich, to develop the various plant collections native to the 600 acre site. This was the humble and serendipitous beginning of today's grand Gardens. It is the case with most botanic gardens that education and learning are central to their function, yet at The Huntington, these values have only ever existed on an almost inadvertent level. Entertainment, not education, has long been the primary objective of Huntington: it was only in 2005 that the Rose Hills Foundation Conservatory for Botanical Science was established on the site.

However, the centre's targets to teach have always been apparent, existing on a more social—rather than academic—level. Not just for scholars, not just for connoisseurs. The Huntington was established for the people: to provide a portal through which the public could access the most distinguished examples of American and Western culture. This is still integral to the centre's goals, despite the fact that now over 1,700 academic researchers use the facilities here each year.

The Huntington's heart is undoubtedly the Library. Housing more than five million manuscripts—some on public display, but most archived in the collection—the Library holds a very special collection of Western literary and academic texts. Highlights include the Ellesmere manuscript of Chaucer's *The Canterbury Tales* a Gutenberg Bible and original correspondence from all four stalwarts of American political history—Washington, Jefferson, Franklin and Lincoln. There is also a vast collection of early editions of Shakespeare's works. This is a neat tie-in with one section of the Gardens, where there is a display devoted to the many plants and flowers mentioned in Shakespeare's plays, each accompanied with a plaque bearing the relevant verse. For example, a pomegranate tree is placed alongside a panel reading "It was the nightingale, and not the lark, That pierc'd the fearful hollow of thine ear, Nightly she sings on yond pomegranate-tree" from *Romeo and Juliet*. This one instance exemplifies the curators' ongoing desire to encourage an interaction between the various resources to be found on the Huntington estate.

Adjacent to the Shakespeare Garden is the Virginia Steele Scott Gallery, home to numerous icons of modern American painting including works by Edward Hopper, Mary Cassatt and John Singer Sargent. This is one of three galleries built on site—others are devoted to historical British painting, European Renaissance works, and another collection of American masterpieces. Being so extensive and so dense, the art collection at Huntington has been divided chronologically and geographically, and this curatorial scheme is identical to the one employed in the Botanical Gardens.

Besides the conventional components found in most botanic gardens—such as the Desert Garden and Rose Gardens—the horticultural curators at Huntington have taken an inventive stance when dealing with their array of 14,000 specimens. There are geographical sections here not commonly found in botanic gardens, such as the Australian Garden, housing 150 species of eucalyptus, as well as acacias, cycads, and melaleucas. There are also plans to develop a Chinese Garden (as opposed to the more common Japanese version), set to spread over 12 acres. This will include pavilions, stone bridges, and a one-acre lake!

One of the most important highlights of the grounds is certainly the North Vista Garden. As with the Shakespeare Garden, this area

quotes the Huntington art collection, specifically the Renaissance works of the Arabella Huntington Memorial Collection (housed in the west wing of the Library). Some of the seventeenth century sculptural works are on display at the North Vista, with rows of Italian statues copied from historic botanic gardens such as Italy's Padua. The Gardens themselves frames a view of the San Gabriel Mountains, with tall columns of Fountain Palms (*Livistona australis*) mounting the panorama.

tag: Aeonium Cyclops in the Desert Garden. above: A variety of cacti in the Desert Garden. below: Created in 1912, the Japanese Garden features a small lake spanned by a moonbridge, a traditional house, and trellises of wisteria that bloom in early Spring. all photographs: The Huntington Library, Art Collections, and Botanical Gardens

UNITED STATES
MISSOURI BOTANICAL GARDENS, ST LOUIS

One day in the spring of 1819, 18 year old Henry Shaw, an Englishman recently landed in the river town of St Louis on the edge of the American wilderness, took a half-day journey on horseback out of town. Riding westward through marshy ground, past sinkholes and Indian burial mounds, he came at last to a narrow path cutting through brush, and found himself on elevated ground overlooking a prairie. "Uncultivated", he recorded, "without trees or fences, but covered with tall luxuriant grass, undulated by the gentle breeze of spring."

If ever a man loved a piece of ground, it was Shaw. Shaw's fortunes grew, he resolved to return something to his adopted city, and 40 years after his arrival in St Louis, he opened a botanic garden on the land he so loved for the benefit of the city's residents. This garden became the Missouri Botanical Garden. One of the oldest botanic gardens in the United States, it is outstanding not only in the excellence of its displays, but also in the richness of its architectural heritage and the importance of its botanical research.

Shaw also had a strong interest in botany and gardening, fostered by early years at Mill Hill School, located on the former estate of English botanist Peter Collinson. It was not surprising, therefore, that one of his trips to England inspired him to give the people of St Louis a garden like the great gardens and estates of Europe. Shortly after 1851 Shaw began development of a ten acre site near his country home. His unusual gesture presaged the age of American philanthropy and the creation of the great American public parks by several decades.

The news that Henry Shaw was building a botanic garden reached Dr George Engelmann, a German physician-botanist who had come to the United States several decades earlier. Engelmann, one of the great early American botanists, suggested that the garden be more than a public park, and suggested it become more involved with scientific work, like the great botanical institutions of Europe. With the assistance of Harvard botanist Asa Gray and Sir William Hooker, director of the Royal Botanic Gardens at Kew, Engelmann persuaded Shaw to include a herbarium and a library in his garden.

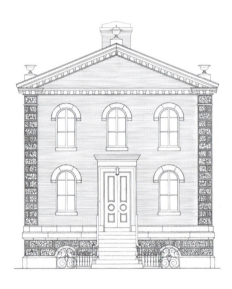

SOUTH ELEVATION

NORTH ELEVATION

Shaw in turn encouraged Engelmann to buy specimens and books in Europe. The Missouri Botanical Gardens opened to the public in 1859 and began to flourish in the European tradition of horticultural display, combined with education and conservation. Today, the Missouri Botanical Gardens covers 79 acres and operates the world's most active tropical botany research programme in the country.

The Garden's 79 acres of splendid horticultural displays include the vibrant tropical rainforest that thrives inside the conservatory. The Japanese Garden covers 14 acres, making it one of the largest Japanese strolling gardens in North America. Other outstanding displays include Chinese, English, German and Victorian gardens. Over 4,000 trees live on the grounds, including some rare and unusual varieties and a few stately specimens dating back to the nineteenth century, when Henry Shaw planted them. The William T Kemper Center for Home Gardening, the nation's most comprehensive resource centre for gardening information, includes 23 residential scale demonstration gardens.

Text by Joseph M Schuster (except last two paragraphs). Reprinted by permission from *Garden*, January–February 1983, the publication of The Garden Society, a Division of The New York Botanical Garden. tag: Spring flowers in bloom. previous pages top and bottom left: Plans of the Gardens' main building elevations. previous pages right: A wake for Henry Shaw, founder of the Gardens' Cactus Society. left: View of the Drum Bridge in Autumn. right: View of the Missouri Climatron with the Waterlily Pond in the foreground.

The New York Botanical Garden is a vast, 250 acre collection of flowers, flowing streams and peaceful wetlands. It is also a premier research institution with 48 gardens and plant laboratories that attract scientists and researchers from around the world. The Garden was founded in 1891 by Columbia University botanist Nathaniel Lord Britton, and at its heart lie 40 acres of virgin woodland, the last remaining original forest that covered New York City before settlers arrived in the early seventeenth century.

Towards the rear of the Garden is the Peggy Rockefeller Memorial Rose Garden originally built in 1916 and redeveloped in 1987.

This garden is closed every winter and re-opens in May, when visitors can climb to the top of the stone steps and look down on a collection of 2,700 rose plants.

Other notable aspects of the Garden include a three acre Japanese Rock Garden that features thousands of alpine flowers and woodland plants, and a 37 acre Arboretum with pine, spruce and fir trees from around the world. The Enid A Haupt Conservatory is a Victorian-style greenhouse built in 1902. It is home to the permanent exhibition A World of Plants, which displays 11 distinct natural habitats, including two types of rainforests and deserts

of the Americas and Africa, as well as an extensive collection of carnivorous and aquatic plants.

Since 2006, the garden has contained the Pfizer Plant Research Laboratory (PPRL), a research institution that focuses plant genomics—the study of genetics in plant development. This institution was set up thanks to the already existing research base that the Garden had invested in. Since 1891, 2,000 plant-collecting missions resulted in millions of specimens that have all been carefully stored, their DNA providing crucial information from which to work.

tag: A variety of Orchids, including *Phalaenopsis, Dendrobiums, Renanthera* and *Cymbidium*, hang the Conservatory display. previous pages: The Peggy Rocke feller Rose Garden, designed by Beatrix Farrand in 1916. above left: The Upland Tropical Rain Forest Gallery, located in the Enid A Haupt Conservatory. above right: A boxwood knot parterre, which forms the structure for the Herb garden. below: Amazon waterlilies (*Victoria Amazonia*), and Lotuses (*Nymphea*), which cultivate in the ponds next to the Enid A Haupt Conservatory. opposite: An alley of Tulip Trees (*Liriodendron*), which leads to the Library.

UNITED STATES
BROOKLYN BOTANIC GARDEN, NEW YORK

Founded in 1910 on the site of a former ash and rubble dump, Brooklyn Botanic Garden is now one of North America's premier public gardens. A tranquil, 52 acre urban oasis situated in the heart of Brooklyn, New York, it hosts more than 700,000 visitors annually, and features more than 10,000 different kinds of plants from around the world.

The Garden's bonsai collection is considered one of the finest in the world, and is the second largest on public display outside of Japan, containing over 200 cherry trees, comprising some 42 species and cultivars. The flowering of the cherries in April and May is Brooklyn Botanic Garden's most spectacular annual event and forms the centrepiece of its Sakura Matsuri—a weekend long cultural festival many consider to be New York City's 'rite of spring'.

Majestic and award-winning Cranford Rose Garden, one of the largest and most revered rose gardens in the country, was opened in 1928 and features more than 5,000 bushes of nearly 1,200 rose varieties, including wild species, old garden roses, and many modern cultivars. The Fragrance Garden, dating from 1955, was designed for the visually-impaired, and is one of Brooklyn Botanic Garden's most popular exhibits.

Teaching children and adults about plants and gardening and instilling them with a love of nature are central to Brooklyn Botanic Garden's mission, and it was the first botanic garden in the world to establish a garden specifically for children, in 1914. Every year, 800 youngsters from city neighborhoods plant crops and flowers and harvest fresh produce in the Children's Garden under the guidance of expert instructors. For older children, lessons in science and ecology enrich the gardening classes.

With its wildlife meadow, sensory planting beds, and children-sized exhibits, the smaller and more intimate Discovery Garden encourages toddlers and their families to interact with the natural environment and explore plant life. Hands-on weekend workshops enhance the Discovery Garden experience. The Brooklyn Botanic

Garden also has a wide array of continuing education classes for adults, including certificate courses in horticulture and floral design.

Brooklyn Botanic Garden prides itself on its strong connection to local communities in the borough of Brooklyn and to the New York Metropolitan Region in general. It is involved in numerous partnerships with sister cultural institutions, city governmental departments, and civic groups to promote environmental education, local greening initiatives, and an appreciation for plants and gardening. A new and historic endeavour has been the

Garden's key role in the creation and operation of the Brooklyn Academy of Science and the Environment (BASE), the borough's first high school dedicated to field-based environmental studies.

Scientific research at Brooklyn Botanic Garden is also regional in scope. The core of the effort is the award-winning New York Metropolitan Flora Project (NYMF). Begun in 1989, NYMF is the Garden's long-term project to inventory all plants growing within a 50 mile radius of New York City. Botanists working on the project have already identified several new, endangered, as well as potentially invasive species and have rediscovered species thought to be extinct. Their important work is being used by other scientists, urban planners, land managers, and policy makers to preserve and restore the metropolitan environment.

tag: The Tree Peony Collection in summer. opposite: The Palm House with the Lily Pool Terrace in the foreground. above: The Desert Pavilion, situated in the Steinhardt Conservatory.

THE SANTA BARBARA BOTANIC GARDEN

The Santa Barbara Botanic Garden is the product of a collaboration between the Carnegie Institution and the Santa Barbara Museum of Natural History, and was conceived in 1925 by plant ecologist Dr Frederic Clements as a garden "reaching from the sea to the crest of the mountains, connected by a drive lined with trees, shrubs and flowers from all parts of the earth". The idea came into fruition in instalments. In 1926 a stretch of 13 acres in Mission Canyon was donated by local philanthropist Anna Dorinda Blaksley Bliss. A further stretch of land was donated in 1927, and in 1932, the final piece was added to create the 78 acre site that the Garden occupies today.

The aim of the founders was to create a garden that would unite the aesthetic, educational and scientific. Based on concepts developed by Clements, the Garden was initially laid out in various plant communities, such as chaparral, desert, and prairie, with an emphasis on plants from the Pacific slope of North America. Experimental groupings of significant genera such as *Ceanothus* and *Eriogonum* (buckwheat) were also displayed for horticultural research and to educate the public. By 1936 this emphasis had narrowed to plants native to the state of California and now includes north western Baja California and south western Oregon, which are both part of the California Floristic Province.

California is one of earth's 25 biodiversity hotspots—areas with unusually high numbers of species. Among the 5,800 native plant species occurring in California, 1,700 have such small populations that they face extinction in the next 50 years. Despite an extensive history of scientific study, there is still scarce knowledge of the role most California native plants play in natural landscapes and how they contribute to human society. The Garden is committed to researching these questions and conserving the rarest plant species of the Central Coast Bioregion, which encompasses Monterey and San Benito Counties south to Ventura County, and includes California's Channel Islands. The conservation team at the Garden collects seeds and cuttings from endangered plant species, managing re-introduction efforts and helping to sustain the surviving populations.

In 2003, a remarkable partnership between the Garden, the community, and the County of Santa Barbara Historic Landmarks Advisory Commission resulted in the adoption of a resolution granting County Historic Landmark status to 23 of the Garden's 78 acres—including the Aqueduct, Indian Steps, Entry Steps, the Information Kiosk, the Caretaker's Cottage, Blaksely Library, and the Campbell Bridge.

tag: The vibrant California poppy (*Eschscholzia californica*). below: Giant Coreopsis (*Coreopsis gigantean*), located in the Gardens' Ceanothus Exhibit. all photographs: Blaksley Library Photo Archives. opposite: Built in 1807, by Native Americans under the authority of the Franciscan padres, the dam and aqueduct are part of the original Mission Waterworks system that brought water to the Santa Barbara Mission.

UNITED STATES
UNITED STATES BOTANIC GARDEN, WASHINGTON, DC

The oldest botanic garden in North America, the United States Botanic Garden, is located on the Nation's Mall in Washington, DC. It is free to the public and open all days of the year.

The Garden was established in 1820 by the Colombian Institute for the Promotion of Arts and Sciences. It was a small-scale project that struggled to survive, and was closed down in 1837. In 1842, the idea of a national botanic garden was reignited when the United States Exploring Expedition to the South Seas (the Wilkes Expedition) brought a collection of living plants from around the globe to the Capitol. Initially placed in a specially constructed greenhouse behind the Old Patent Office Building, the plants were moved in late 1850 into a new structure on the site previously occupied by the Columbian Institute's Garden.

The Garden was moved to its present location in 1933, and preserves many of its original features. In recent years the Garden has been worked on significantly. The most major change has been the renovation of the Conservatory to include a catwalk bridge hung from its ceiling, so that visitors can walk through the treetops inside it. The Conservatory is home to some of the Garden's oldest plants, and many plants from the Wilkes expedition still remain in the collection today.

Just south of the Conservatory is Bartholdi Park, named after Frederic Auguste Barholdi, who designed the Statue of Liberty. This section of the Garden is given over to a more contemporary collection of plants than the Conservatory. Designed as a triangle with a magnificent fountain at its centre, this garden is continually updated, and showcases innovative plant combinations. Near this park is an 25,603.2 metre-squared labyrinth of climate-controlled zones, known as the Production Facility. This is the largest greenhouse complex supporting a public garden in the United States. Constructed in 1994, the Production Facility serves as a nursery for Garden and other gardens in the area.

The most recent addition to the United States Botanic Garden is the National Garden, opened in 2006 and featuring a garden of Atlantic and coastal plain plants, a rose garden, a butterfly garden and a water garden.

tag: A summer bed in bloom with the Historic Frederic Auguste Bartholdi Fountain and glass dome of the Conservatory in the background. above: The Capitol and central glass dome of the Conservatory, as viewed from the new National Garden. opposite: Central path and stairs in the Jungle Section of the Conservatory.

Biographies,
Bibliography,
Acknowledgements
and Index

Biographies

Brian Johnson is an education officer for Botanic Gardens Conservation International (BGCI) based in both London and New York, which is dedicated to plant conservation and sustainable development education for both children and adults. He has published numerous articles on the importance of environmental education and was a senior faculty member at the Audubon Expedition Institute at Lesley University in Cambridge, Massachusetts from 1998 to 2003 before serving as the Director for Programs at Prospect Park Audubon Centre in Brooklyn, New York from 2003–2004.

Fabio Garbari is an expert on Mediterranean flora and is professor of Systematic Botany at Pisa University. He co-authored *Gardens of Simples*, 2002, a text on the history, people, and roles of Pisa Botanical Gardens over the centuries. Much of his research is dedicated to discovering the relationship between art and science, with particular focus on sixteenth through eighteenth century botanical iconography. He is also director of the Botanical Gardens and Museum of the Department of Biology at the University of Pisa and president of Si MA. (Museum and Collections System of the Pisan Athenaeum).

Gerda van Uffelen is the collection manager of the Hortus Botanicus at Leiden and is considered an expert on ferns. She designed the new systematic garden and is actively researching the early years of the unique garden.

Gregory Long has been leading cultural institutions in New York City for over three decades. He became president and chief executive officer of the New York Botanical Gardens in 1989 after seven years with the New York Public Library. He has also served on the Mayors Advisory Commission for Cultural Affairs, and has brought unprecedented revitalisation and growth to the Garden.

Holly H Shimizu has spent her career promoting public horticulture in various areas of the world and is now executive director of the United States Botanic Garden in Washington, DC. Shimizu works to educate and inspire people about the critical importance of plants in our daily lives.

John Parker is a professor of Plant Cytogenetics at the University of Cambridge where he also works as the director of the University Botanic Garden and curator of the Herbarium. A former trustee of the Royal Botanic Gardens, Kew, and council member of the Royal Horticultural Society, he is now the director of the Natural Institute for Agricultural Botany and an honorary research fellow at the Natural History Museum in London. His research primarily focuses on the genetics of plant populations as well as the origins of the modern theory of evolution.

Margaret Stevens is an award-winning botanical artist, focusing her palette on scenes of flora and fauna. Her work has been seen in some of the UK's leading botanical publications and featured in *This England* magazine for over 16 years. She has been awarded 13 medals by the Royal Horticultural Society for her work including the Gold and the Silver Gilt Lindley medal for work of special education merit and she became a founder of the Society of Botanical Artists (SBA) in 1981. She is now president of the SBA and is course director for the society's distance learning diploma course in botanical painting.

Mike Maunder is a fourth-generation horticulturist. He works as the executive director of the Fairchild Tropical Botanic Garden in Coral Gables Florida and has degrees in plant taxonomy and conservation genetics. He also serves as chair of the World Conservation Union's Plant Conservation Committee and is director of the American Public Garden Association (APGA).

Nina Antonetti is an experienced architectural historian and researcher, who works as an assistant professor in the new landscape studies programme at Smith College in Northampton, Massachusetts. She has held research positions at the Centre for Advanced Study in the Visual Arts, National Gallery of Art, Washington, DC, and at the Victoria and Albert Museum in London.

Rosie Atkins has dedicated her life to writing on gardening. After working on the *Sunday Times* newspaper from 1968 to 1982, she went on to become the gardening correspondent of *TODAY* before launching *Gardens Illustrated* magazine in 1993. She left *Gardens Illustrated* in 2002 and became curator of the Chelsea Physic Garden in London. She is now a Fellow of the Linnean Society, serves on the Horticultural Board and the Woody Plant Committee of the Royal Horticultural Society and is a trustee of Gardening for the Disabled.

Bibliography

Baker, K, *Tempering the Elements: Botanic Gardens and the Search for Paradise*, Geneva: The 23rd Confererence on Passive and Low Energy Architecture, 2006.

Bannavti, OT, "Education Programmes at Limbe Botanic Garden, Cameroon: Model solutions", *Roots*, July 1995.

Bartholomew, James, *The Magic of Kew*, Chicago: New Amsterdam Books, 1988.

Berenson, Richard J and Neil De mause, *The Complete Guidebook to Prospect Park and the Brooklyn Botanic Gardens*, Santa Monica: Silver Lining, 2001.

Blumin, SM, "Driven to the City: Urbanization and Industrialization in the Nineteenth Century", *OAH Magazine of History*, vol. 20, no. 3, 2006, pp. 47–53.

Bramwell, David, O Hamann, V Heywood, and H Synge, *Botanic Gardens and the World: Conservation Strategy*, Oxford: Academic Press, 1987.

Brockway, Lucille H, *Science and Colonial Expansion: The Role of the British Royal Botanic Gardens*, London: Yale University Press, 2002.

Cappelletti, EM, "The Botanic Garden of the University of Padua 1545–1995", *BGCNews*, vol. 2, no. 4, 1994, retrieved 13 November 2006, from *BGCNews* Journal Archive database.

Chengyih, W, and T Fengqin, eds., *The Blossoming Botanical Gardens of the Chinese Academy of Sciences*, Beijing: Science Press, 1997.

Chihuly, Dale and Todd Alden, *Chihuly at the Royal Botanic Gardens, Kew*, London: Portland Press, 2005.

Comito, T, "Renaissance Gardens and the Discovery of Paradise", *Journal of the History of Ideas*, vol. 32, no. 4, 1971, pp. 483–506.

Curtis, Eric, *The Story of Glasgow Botanic Gardens*, Glendaruel: Argyll Publishing, 2006.

Dejun, Y, ed., *The Botanical Gardens of China*, Beijing: Science Press, 1983.

Desmond, Ray, *History of Kew*, London: The Harvill Press, 1995.

Desmond, Ray and Sir Ghillean Prance, *Kew: The History of the Royal Botanic Gardens*, London: The Harvill Press, 1998.

Drewitt, FD, *The Romance of the Apothecaries' Garden at Chelsea* (third edition), Cambridge: Cambridge University Press, 1928.

Dunlop, Eric, *The Story of the Dunedin Botanic Garden*, Dunedin: The friends of the Dunedin Botanic Garden, 2002.

Eyre, Alan, *The Botanic Gardens of Jamaica*, London: Andre Deutsch, 1966.

The Flora of the Cayman Islands, London: The Stationery Office Books, 1984.

Furse-Roberts, James, *Botanic Garden Creation: the Feasibility and Design of New British Collections*, Reading: University of Reading, 2005.

Gager, CS, "Botanic Gardens of the World: Materials for a History", *Brooklyn Botanic Garden Record*, vol. XXVI, no. 3, 1937 pp. 149–353.

Garbari, Gianni and Fabio Bedini, eds., *The 400 Years of the Pisa Botanic Garden: Is Its Past the Key to Its Future?*, Pisa: Dipartimento Di Science Botaniche, Universita Di Pisa, 1991.

Gardens in One: the Royal Botanic Garden Edinburgh, London: The Stationery Office Books, 1992.

Gilbert, Lionel, *The Royal Botanic Gardens, Sydney: A History 1816–1985*, Oxford: Oxford University Press, 1988.

Global Strategy for Plant Conservation, Montreal: Secretariat of the Convention on Biological Diversity, 2002.

Graham, Anne Marie, et al, *A Garden for All Seasons: Artist's View of the Royal Botanic Gardens*, Sydney: Craftsman House, 1998.

Hallett, S, "World's First Botanical Garden has Roots in Medicine", *Canadian Medical Association Journal*, vol. 175, no. 2, 2006, p. 177.

Hardiman, Lucy and C Colston Burrell, *Intimate Gardens (Brooklyn Botanic Garden All-Region Guide)*, New York: Brooklyn Botanic Garden, 2005.

Hepper, F Nigel, *Bible Plants at Kew*, London: The Stationery Office Books, 1981.

Hepper, F Nigel, *Kew: The Royal Botanical Gardens—Gardens for Science and Pleasure*, Owings Mill: Stemmer House, 1997.

Heywood, VH and Peter Wyse, *Tropical Botanic Gardens: Their Role in Conservation and Development*, Oxford: Academic Press, 1991.

Hill, AW, "The History and Function of Botanic Gardens", *Annals of the Missouri Botanical Garden*, vol. 2, February 1915, pp. 185–240.

Hobson, Diana, Keiko Mukaide and Craig Mackay, *Elemental Traces*, Edinburgh: Royal Botanic Gardens, 2000.

Hyams, E, and W MacQuitty, *Great Botanical Gardens of the World*, New York: The Macmillan Company, 1969.

Kerner, A, *Botanic Gardens: Their Functions in the Past, the Present and the Future*, Unpublished Manuscript, 1874.

Lodari, Carola and Marco Capovilla, *Villa Taranto: Captain McEacharn's Garden (Archives of Botanic and Garden Studies)*, Turin, 1999.

MacPhail, I, *Hortus Botanicus: The Botanic Garden and the Book: Fifty Books from the Sterling Morton Library Exhibited at the Newberry Library for the Fiftieth Anniversary of the Morton Arboretum*, Chicago: The Morton Arboretum, 1972.

Marinelli, Janet, *Brooklyn Botanic Garden Gardener's Desk Reference*, New York: Henry Holt, 1998.

Marinelli, Janet, *Plant (Royal Botanic Gardens Kew)*, London: Dorling Kindersley Publishers Ltd, 2004.

Martin, Tovah and Barbara B Pesch, *Greenhouses and Garden Rooms (Plants and Gardens, Brooklyn Botanic Garden Record)*, New York: Brooklyn Botanic Garden, 1999.

Martins, David and Andrew Guthrie, *Queen Elizabeth II Botanic Park: One With Nature*, Marceline: Walsworth Publishing, 2002.
McCracken, DP, *Gardens of Empire: Botanical Institutions of the Victorian British Empire*, London: Leicester University Press, 1997.

Morris, Deborah and Greg Elms, *The Royal Botanic Gardens, Melbourne: A Life and Times*, Melbourne: Allen and Unwin, 2001.

Myrie, S and E Arnone, *Connecting with Teens: Strategies for Engaging Youth in Botanic Gardens*, Oxford: The Nature of Success: Success for Nature, 2006.

Pegler, David, *Genus Lentinus: A World Monograph*, London: The Stationery Office Books, 1984.

Pharaoh's Flowers: Botanical Treasures of Tutankhamen, London: The Stationery Office Books, 1999.

Prest, John, *The Garden of Eden: The Botanic Garden and the Re-creation of Paradise*, London: Yale University Press, 1982.

Pridmore, Jay and Arthur Lazar, *A Garden for All Seasons: the Chicago Botanic Garden*, Chicago: Chicago Horticultural Society, 1991.

Reeds, KM, *Botany in Medieval and Renaissance Universities*, New York: Garland Publishing, 1991.

Sclater, Andrew, ed. *The National Botanic Garden of Wales*, London: HarperCollins, 2000.

Soderstrom, M, *Recreating Eden: A Natural History of Botanical Gardens*, Vehicule Press, 2001

Stearn, WT, "The Chelsea Physic Garden 1673–1973: Three Centuries of Triumph in Crises. A Tercentenary Address", *Garden History*, vol. 3, no. 2, 1975, pp. 68–73.

Stern, WL, "The Uses of Botany, With Special Reference to the 18th Century", *Taxon*, vol. 42, no. 4, pp. 773–779.

Thacker, C, *The History of Gardens*, London: Croom Helm, 1979.

Tobey, GB, *A History of Landscape Architecture: The Relationship of People to Environment*, New York: Elsevier North Holland, 1979

Waylen, Kerry, *Botanic Gardens: Using Biodiversity to Improve Human Well-Being*, Richmond: Botanic Gardens Conservation International, 2006.

Whittle, T, *The Plant Hunters*, New York: PAJ Publications, 1988.

Wilson, Julia, *Education for Sustainable Development: Guidelines for Action in Botanic Gardens*, Richmond: Botanic Gardens Conservation International, 2006.

Wing Yam, Tim, *Orchids of the Singapore Botanic Gardens*, Singapore: Singapore Botanic Gardens, 1995.

Wyse Jackson, PS, and LA Sutherland, *International Agenda for Botanic Gardens in Conservation*, London: Botanic Gardens Conservation International, 2000.

Index

Acknowledgements

Thank you to Brigitte Monem, John Parker and Lorraine Robertson for your guidance and inspiration, and to RBG Kew Gardens Library for your research advice and support. We appreciate the hard work of our researchers and profile writers; Tom Morris, Kasia Behnke, Guy Reder, Shumi Bose, Jill Kovacs, Sally McHugh, Alastair Coe and Peter Kneller.

We are grateful to the Foundation for Landscape Studies for permission to reprint the essay "The Botanical Garden" from *Sitelines: A Journal of Place*, vol. II, no. 1 Fall 2006.

Perhaps most importantly, we are very much indebted to the many Gardens who provided the support necessary to create this book. This volume is dedicated to the work of these increasingly important institutions.

Colophon

© 2007 Black Dog Publishing Limited
All rights reserved

Editor: Nadine Käthe Monem
Assistant Editor: Blanche Craig
Designer: aleatoria

ISBN 10: 1-904772-72-2
ISBN 13: 978-1-904772-72-9

Black Dog Publishing Limited
Unit 4.4 Tea Building
56 Shoreditch High Street
London E1 6JJ

Tel: +44 (0)20 7613 1922
Fax: +44 (0)20 7613 1944
Email: info@blackdogonline.com
www.blackdogonline.com

British Library Cataloguing-in-Publication Data.
A CIP record for this book is available from the British Library.

Every effort has been made to trace the copyright holders, but if any have been inadvertently overlooked the publishers will be pleased to make the necessary arrangements at the first opportunity.

Black Dog Publishing is an environmentally responsible company. Botanic Gardens is printed on Garda Matt Art 170 gsm, an acid-free paper made with cellulose from certified forests, plantations and well managed forests.

architecture art design
fashion history photography
theory and things

www.blackdogonline.com